TABLE OF CONTENTS

T0140622

Meteoric cosmogenic nuclides   43

"Insitu" cosmogenic nuclides   51

Seminars

Theses

Collaborations

Visitors

EDITORIAL

The **Laboratory of Ion Beam Physics** (LIP) is a leading research center for the development of Accelerator Mass Spectrometry (AMS) and a world-class laboratory for the application of Ion Beam Physics in a wide range of fields. It can rely on a broad funding base and can thus serve a large user community from many highly current disciplines. LIP acts as a national competence center for ion beam applications.

This annual report is a brief summary of the achievements accomplished in 2013. It covers the wide range of fields from fundamental research, over operational issues of the laboratory, to the vast variety of exciting applications of our measurement technologies. Naturally, the publications listed in the appendix of this report will provide a much deeper insight into our research.

Certainly, one highlight in 2013 was the general audit of the ETH Zurich Department of Physics that took place in April, where a dedicated subpanel of the review board took special care to evaluate the LIP activities. The very positive evaluation report has encouraged us to proceed with our general strategy. The renewed curatorial agreement provides a solid base for the LIP operation in the near future.

Research projects to improve AMS instrumentation continued. Two new MICADAS projects should further the progress of AMS capabilities. The use of helium as stripper gas has been fully implemented into the MICADAS system that will be delivered to Aix-Marseille University. This system benefits from improved measurement conditions increasing the overall transmission to almost 50 %. The second project will integrate the new permanent magnet technology into a MICADAS system. The University of Uppsala will be the first institution to benefit from the new technology, which should reduce the operational costs of the instrument significantly. **Ionplus**, a new ETH-spinoff company has been founded in February 2013 to commercially produce technical instruments developed at LIP to optimize operational laboratory procedures. At present, automated graphitization systems (AGE III), gas handling interfaces, gas ionization detectors, and other auxiliary tools are on the **Ionplus** product list.

The close connection to our partners at the Paul Scherrer Institut, Empa, Eawag, and other ETH departments has been successfully continued. In particular, the many collaborative projects with the Earth Sciences Department have resulted in a large number of ^{14}C analyses, both with solid and gaseous CO_2 samples.

I am grateful to the excellent LIP scientific, technical and administrative staff, which is responsible for the success of the laboratory. The well maintained technical infrastructure is the platform to support the large variety of application projects LIP members are contributing to. The versatile instrumentation will continue to provide excellent service to external users and contribute significantly to the educational program of ETH. Special thanks go to Peter Kubik who had the editorial responsibility not only for this report but also for the reports of the past four years. He will retire end of 2014 and we wish him all the best for the coming stage in his life.

Hans-Arno Synal

THE TANDEM AMS FACILITY

Operation of the 6 MV TANDEM accelerator

Energy-loss straggling in gases

OPERATION OF THE 6 MV TANDEM ACCELERATOR

Beam time and sample statistics

Scientific and technical staff, Laboratory of Ion Beam Physics

After about two years dominated by a number of tank openings related to the conversion of the 6 MV EN Tandem charging system to the NEC Pelletron system [1], we entered a more stable period during the last half of 2013. As a result, the total number of operation hours decreased in 2013, because less time was required for conditioning or maintenance (Fig. 1). On the other hand, the total time used for AMS measurements increased again to about 900 h, partitioned into about 400 h for ^{10}Be and ^{36}Cl each and about 100 h for ^{26}Al. About 10 % of the time (included in the AMS fraction in Fig. 1) was used for energy-loss straggling measurements. The number of analyzed samples in the materials sciences has increased with less beam time as a result of improved efficiency.

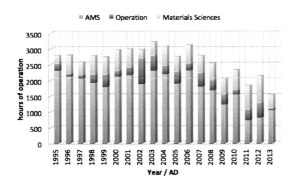

Fig. 1: *Time statistics of the TANDEM operation subdivided into AMS, materials sciences, and service and maintenance activities.*

In 2013, about 620 Samples were analyzed for ^{10}Be, divided into 320 ice core samples, 200 *in-situ* dating samples, 50 meteoric ^{10}Be and some calibration samples.

Ten years of measuring samples from the EPICA project's DML ice core (Fig. 2) resulted in about 8000 ^{10}Be data points in total. These measurements provide a unique record of the ^{10}Be concentration during the last roughly 800,000 years and allow e.g. reconstruction of solar activity in the past [2].

Fig. 2: *Sampling station of Dronning Maud Land (DML) in Antarctica (Photo: H. Fischer, University of Bern).*

150 samples were analyzed for ^{26}Al mainly to complement *in-situ* ^{10}Be data.

In about 400 samples we measured ^{36}Cl; half of them were for *in-situ* dating of rocks, one third were extracted from irradiated material and the rest were water samples from various places.

For the materials sciences beam time increased slightly from 400 to 460 hours and the number of analyses and irradiations from 1000 to 1080. Approximately 75 % of all analyses were Rutherford Backscattering (RBS), followed by heavy ion elastic recoil detection analysis (ERDA, 15 %) and particle induced X-ray emission (PIXE, 5 %). Only 60 to 70 irradiations were performed. However, they account for a major share of the beam time, since irradiations can take much longer per sample than ion beam analyses.

[1] C. Vockenhuber et al., Laboratory of Ion Beam Physics Annual Report (2011) 9

[2] F. Steinhilber et al., Proc. Nat. Acad. Sci. United States America 109 (2012) 5967

ENERGY-LOSS STRAGGLING IN GASES

First measurements using Si beams to test theoretical models

M. Thöni, C. Vockenhuber, R. Gruber, T. Mettler

We performed energy-loss straggling measurements with Si beams over a wide energy range and in various gases (He, N_2, Ne, Ar and Kr) to test theoretical predictions. The Si ions were accelerated with our EN Tandem to 15 - 66 MeV. The energy range up to about 300 MeV will be investigated at the cyclotron of the University of Jyväskylä (Finland) to cover the straggling peak expected at ≈ 100 MeV.

We used the external stripper of the TANDEM AMS beam line as a windowless gas target. It is located after the charge (and energy) selecting ESA and close to the object point of the analyzing magnet. To reach a high enough target thickness, a 70 cm long tube with a diameter of 3 mm was placed inside the existing stripper canal (Fig. 1). This way, a pressure of up to 2 mbar could be achieved in the middle of the tube without tripping the turbo pumps used for differential pumping. A series of small apertures (0.5 and 1 mm) in front and after the stripper reduced contributions from ions scattered on the long tube walls. The whole assembly was carefully aligned.

Fig. 1: *The external stripper modified with a 3 mm diameter tube inside.*

The energy distribution of a ^{28}Si beam exiting the gas target was measured by scanning the beam over a narrow aperture in front of a gas ionization detector (Cl-Det, Fig. 2). Beam fluctuations that could influence the energy distribution measurement during the scan were monitored with ^{30}Si in a second detector with a large energy acceptance (SIMS-Det, Fig. 2). Both Si isotopes were injected sequentially by pulsing the LE magnet chamber similarly to standard AMS measurements.

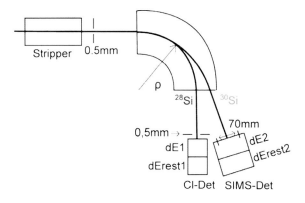

Fig. 2: *Schematic measurement setup with two detectors at the high energy side of EN Tandem.*

Energy-loss straggling is the variance (Ω^2) of the energy-loss distribution and can be calculated from the FWHM of the measured peaks. Contributions from the initial beam energy spread and effects of the measurement setup itself can be subtracted from the profile obtained without gas. Ω^2 should be linearly related to the target thickness (which is proportional to energy loss). The ratio of the slope to the one calculated for Bohr straggling is equal to Ω^2/Ω_{Bohr}^2 (published straggling data are generally normalized to Bohr straggling). This can then be compared to $\Omega_{theory}^2/\Omega_{Bohr}^2$ of various theoretical models.

THE TANDY AMS FACILITY

Activities on the 0.6 MV TANDY in 2013

Improving the TANDY Pelletron charging system

Upgrade of a MICADAS type accelerator to 300 kV

A Monte Carlo gas stripper simulation

^{129}I towards its lower limit

U-236 at low energies

The new beam line at the CEDAD facility

The upgrade of the PKU AMS – The next step

ACTIVITIES ON THE 0.6 MV TANDY IN 2013

Beam time and sample statistics

Scientific and technical staff, Laboratory of Ion Beam Physics

In 2013, the multi-isotope facility TANDY (Fig. 1) was used for routine AMS measurements during more than 2000 hours. About 1800 samples were analyzed for various applications. No major technical modifications were made this year.

Fig. 1: *The compact 0.6 MV TANDY accelerator.*

Most of the time was spent on AMS analyses of ^{10}Be (almost ⅖, Fig. 2) followed by the actinides (about ⅓ of the time) and ^{129}I (about ¼). During 5 % of the operational time the TANDY was used to test and to develop new detectors and detection systems. Finally, about 3 % was spent on the development of ^{26}Al measurements.

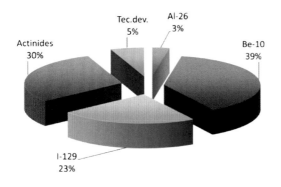

Fig. 2: *Relative distribution of the TANDY operation time for the different radionuclides and activities in 2013.*

More than 800 ^{10}Be samples were analyzed, most of them for ice core studies (Fig. 3). But also lake sediments and samples for in-situ cosmogenic nuclide dating contributed significantly to the total load.

Over 600 ^{129}I samples were measured for different projects. The majority of the ^{129}I analyses were performed within a broad monitoring program in Germany. ^{129}I also was determined in beam dumps, air filters, and ocean water samples.

In total, more than 300 actinide samples were analyzed. A main driver for ^{236}U is our in-house project about the use of anthropogenic nuclides in oceanography. Pu isotopic compositions were measured in samples from the surroundings of Fukushima. New AMS routines for the higher actinides (Np, Am, Cm, and Cf) were developed in collaboration with the Chalk River Labs, Canada.

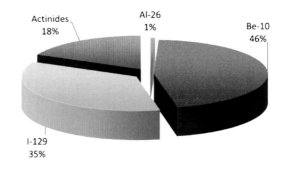

Fig. 3: *Relative distribution of the number of samples measured for the various radionuclides.*

Finally, a few samples were measured for ^{26}Al and first experiments were carried out to test a new absorber method for future AMS analyses of ^{26}Al in the 2+ charge state.

IMPROVING THE TANDY PELLETRON CHARGING SYSTEM

A true physics experiment with 5 multimeters

C. Vockenhuber, M. Christl, S. Maxeiner, A. Müller, B. Ramalingam

The TANDY accelerator has been running very reliably so far. Usually, one tank opening per year is required for maintenance work like cleaning, checking the charging system and refilling the stripper gas bottle in the terminal with He. The most demanding nuclide with respect to the terminal voltage is ^{10}Be, which requires 530 kV. A clear sign for an impending tank opening is an unstable terminal voltage accompanied by a drop in the charging current (Fig. 1).

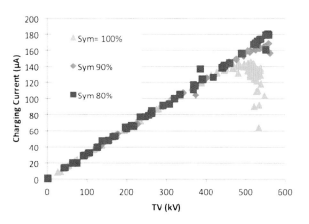

Fig. 1: *Charging current at various suppressor settings (Sym = 100 % e.g. means inductor and suppressors voltages are equal).*

During the 2013 tank opening we investigated this phenomenon in air at lower charging currents (Fig. 2). The charging current drop was caused by discharges between the chain and the suppressor electrode (Fig. 3). This way part of the negative current from the terminal could bypass the measuring point for the charging current. Over time a layer of SF_6 discharge products forms inside the electrode reducing the gap even further with the result that the effect starts at even lower terminal voltages. At a certain point removing this extra layer by cleaning is required to ensure stable operation above 500 kV.

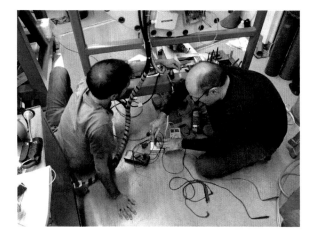

Fig. 2: *Marcus Christl and Sascha Maxeiner during the in-air testing of the charging system.*

During the tests we regulated inductor and suppressor electrode voltages independently and found that the sparking was reduced when the suppressor voltage was lowered to about 80% of the inductor voltage without impairing the main functionality of the suppressor, i.e. suppressing sparking between the chain and the wheel.

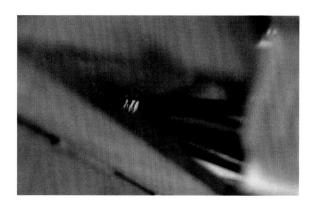

Fig. 3: *Sparking between the chain and the suppressor electrodes on the ground side.*

The clear improvement is seen in the charging current plot (Fig. 1). At 80 % the usual drop of the charging current starting around 500 kV is now gone.

UPGRADE OF A MICADAS TYPE ACCELERATOR TO 300 KV

First steps towards a compact, multi-isotope AMS system

S. Maxeiner, A. Herrmann, H.-A. Synal, M. Christl

With the TANDY AMS system, routine actinide isotopic ratio measurements are made at ion energies of 1.3 MeV, which corresponds to a terminal voltage of 320 kV. A recent study suggests that accelerating voltages as low as 200 kV could be sufficient [1]. The TANDY accelerating voltage is generated by a Pelletron system, which can charge the terminal up to 0.6 MV under SF_6 insulation.

The MICADAS facilities on the other hand are equipped with a vacuum insulated acceleration terminal with a maximum voltage of 200 kV for radiocarbon measurements. This type of accelerator is more compact and easier to maintain than a Pelletron system. A TANDY type spectrometer combined with the MICADAS type accelerator could benefit from the advantages of both systems. To investigate the feasibility of such a system, a 300 kV power supply was connected to a slightly modified stand-alone MICADAS accelerator chamber (Fig. 1).

Fig. 1: *A 300 kV power supply connected to a modified MICADAS accelerator chamber. The chamber does not contain a stripper gas system yet.*

The terminal was carefully conditioned - with the chamber vacuum of 2×10^{-8} mbar - to the maximal voltage of 300 kV over a period of 9 days (Fig. 2). Up to 230 kV, charging currents were less than 30 µA. Above 230 kV, higher charging current of up to 70 µA were needed to condition the chamber. Apart from this, no further problems occurred and the maximum voltage was reached relatively quickly.

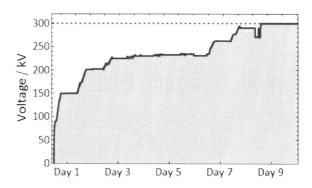

Fig. 2: *Terminal voltage versus time during the conditioning process.*

The high voltage broke down several days later due to a defective section in the supply cable. After the affected section was removed the maximum voltage was reached once again.

The next step, planned for 2014, will be the conditioning of a fully equipped chamber including a He feeding stripper gas capillary operating at a pressure of up to 50 bar. If this step is successful, we plan to temporarily replace the Pelletron of the TANDY system with the new MICADAS chamber to investigate the performance of the prototype system for actinide measurements.

[1]   S. Maxeiner et al., Laboratory of Ion Beam Physics Annual Report (2013) 17

A MONTE CARLO GAS STRIPPER SIMULATION

Simulation of gas flow and projectile scattering inside the stripper tube

S. Maxeiner, M. Suter, M. Christl, H.-A. Synal

The interaction of the ion beam with the stripper medium is of crucial importance for AMS measurements. But also most beam losses due to scattering occur here. For stable measurement conditions and high beam transport efficiency, it is important to reduce these losses by carefully designing the stripper geometry. The present design of the MICADAS gas stripper consists of several, millimeter diameter tube segments. The diameter increases towards both stripper ends to accommodate for the beam divergence (Fig. 1). Gas is flowing from the inlet in the innermost segment to the ends and expands into the surrounding accelerator chamber. To reduce unwanted machine background, this gas flow needs to be minimized while maintaining the gas density needed for molecule disassociation and charge exchange.

their trajectory and scattering processes with the nuclei of the stripper gas atoms [1]. The spatial distribution of the nuclei is calculated from the gas flow and tube segment geometries. Projectiles scattering with the tube walls are considered to be lost.

With this new tool it is now possible to optimize a gas stripper design for optimal transmission. Tube geometries can be investigated for different ion energies, gas types, phase space volumes and beam focusing. Applying the program to the different gas strippers used in our laboratory shows good agreement with measured transmission values.

The simulation data can be used as input for ion optical calculations. For example, the energy dispersion of the high energy magnet leads to a (asymmetrically) broadened ion beam. This can be measured with a Faraday cup. Modeled and measured beam intensities show very good agreement for ^{238}U (Fig. 2).

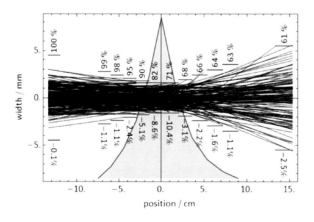

Fig. 1: *Monte Carlo simulation for 200 ions passing through the gas stripper. Scattering losses are indicated by the percentage values (bottom). The cumulative transmission is given on top. The calculated (normalized) gas density is shown by the blue curve.*

To quantify these effects, a Monte Carlo (MC) program was written which generates single ions within a given phase space and simulates

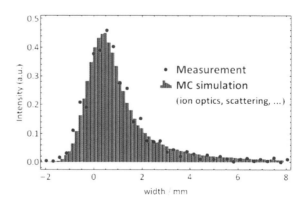

Fig. 2: *Normalized beam profiles of $^{238}U^{3+}$ after the HE magnet at 350 keV (blue: measurements with Faraday cup, red: MC simulation of nuclear scattering in He and subsequent beam transport through the accelerator and the HE magnet).*

[1] G. Amsel et al., Nucl. Instr. & Meth. B 201 (2003) 325

^{129}I TOWARDS ITS LOWER LIMIT

Understanding cross contamination issues for low ^{129}I samples

C. Vockenhuber

Low-energy AMS is well suited for measurements of ^{129}I because the stable isobar ^{129}Xe does not form negative ions, thus high ion energies are not required for isobar separation. We run the TANDY at 300 kV and select charge state 2+ with a transmission of >50 %. In contrast to many other AMS nuclides, ^{129}I readily forms negative ions and the overall detection efficiency is high. The challenges lie more with the ion source where cross contamination can be quite severe due to the volatile nature of iodine. This is particularly important when analyzing samples that are influenced by anthropogenic sources because the isotopic ratios can then span several orders of magnitude.

We have investigated cross contamination at various ion source settings using Woodward (WW) blank samples (Fig. 1) and dummy cathodes (without any iodine). Cross contamination is low when running the source with a low Cs sputter energy (3 keV) and low reservoir temperature. Typical ^{127}I$^-$ currents from mg samples are then 1 - 3 µA.

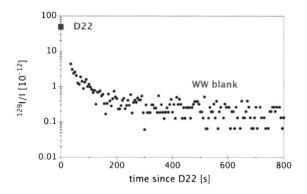

Fig. 1: *^{129}I counting rates of a WW blank sample measured directly after tuning with a standard sample (D22 with ^{129}I/I = 5x10^{-11}).*

Even under these special running conditions samples with high ^{129}I/I ratios (>5x10^{-11}) can

influence the measurements of lower ratio samples. We therefore operate in the following mode: after a first cleaning and check-out pass we measure the samples with low ratios. Once these measurements are finished, the high ratio samples are measured with shorter cycle times and further reduced Cs temperature.

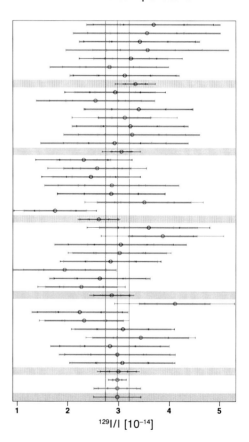

Fig. 2: *Results of five WW blank samples measured together with a low-ratio secondary standard with ^{129}I/I=5x10^{-12}.*

Very low ratio samples (^{129}I/I<10^{-13}) are still influenced by high ratio samples in the wheel even when they are not sputtered. However, these samples can be well measured when no high ratio samples are present in the wheel (Fig. 2). Under these conditions we obtain for our WW blank material a ^{129}I/I ratio of ≈ 3x10^{-14}.

U-236 AT LOW ENERGIES

Detection of ^{236}U with the TANDY set to 200 kV

S. Maxeiner, S. Haas, M. Christl, J. Lachner

Routine ^{236}U measurements at the TANDY AMS facility are usually made at a terminal voltage of 320 kV corresponding to an ion energy of 1.3 MeV (UO$^-$ into U^{3+}, [1]). This is the maximal ion energy for mass 238 ions in charge state 3+ that can pass through the high energy magnets. Despite scattering losses of $\approx$ 15 % estimated with Monte Carlo (MC) methods [2], the maximal system transmission (UO$^-$ into U^{3+}) is around 40 % at these energies.

To investigate the possibility of measurements at lower energies, tests at a terminal voltage (U_{term}) of 200 kV (0.8 MeV detection energy) were performed during a semester thesis [3].

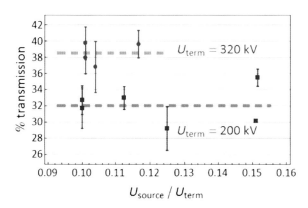

Fig. 1: *Tandy transmission (UO$^-$ into U^{3+}) at stripping energies of 352 keV (top, blue) and 220 keV (bottom, red). Beam focusing into the stripper depends on the ratio U_{source}/U_{term}, which therefore affects the magnitude of scattering losses (U_{source} is 32 kV for U_{term}=320 kV and 20 kV for U_{term}=200 kV).*

The results show a lower transmission of around 32 % (Fig. 1) due to the larger phase space volume of the ion beam at the lower energy, which leads to larger scattering losses in the stripper of around 22 % (estimated by the same MC method). Furthermore, the stripping yield for the 3+ charge state is slightly less at lower ion velocities.

To demonstrate our capability of isotopic ratio measurements at low energies, two different but well-known materials were analyzed for their ^{236}U/^{238}U ratios. The results were normalized to the mean values measured at 320 kV (Fig. 2).

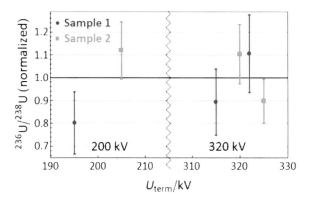

Fig. 2: *^{236}U/^{238}U ratio (normalized to the mean values of measurements at 320 kV) of two different sample types measured at 1.3 MeV and 0.8 MeV detection energy.*

The measured (normalized) ratios are in good agreement and demonstrate the possibility to measure uranium at energies as low as 0.8 MeV, corresponding to a terminal voltage of 200 kV. These energies can also be reached with a MICADAS type accelerator. The somewhat lower transmission is not really a serious issue and could probably be increased with a properly designed terminal stripper canal [2].

[1]   M. Christl et al., Nucl. Instr. & Meth. B 294 (2013) 29

[2]   S. Maxeiner et al., Laboratory of Ion Beam Physics Annual Report (2013) 15

[3]   S. Haas, Semester thesis (2013) Laboratory of Ion Beam Physics

THE NEW BEAM LINE AT THE CEDAD FACILITY

First ^{10}Be tests with the new multi-isotope beam line

A.M. Müller, L. Calcagnile[1], G. Quarta[1], H.-A. Synal

In collaboration with the Laboratory of Ion Beam Physics (LIP) a new multi-isotope HE spectrometer was installed recently at the CEDAD AMS facility located at the University of Salento. It consists of a 60° magnet, 90° ESA and a 135° magnet (Fig. 1). The radionuclides are detected with a ΔE-E_{res} gas ionization chamber built at LIP [1].

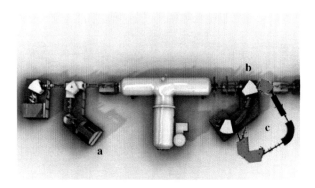

Fig. 1: *Top view of the CEDAD AMS facility [1] with the new multi-isotope spectrometer (c).*

The beam line design was optimized for ^{10}Be measurements with the degrader foil technique. In March 2013, a first ^{10}Be beam was injected into the new beam line and detected in the gas ionization chamber (Fig. 2) with an overall transmission (LE side into the detector) of ≈ 1 %. Optimizing the spectrometer settings increased this value to about 7 - 8 % so far.

The ETH ^{10}Be in-house low level standard BEBL (Fig. 3) was measured at the level of 1.2×10^{-14} in good agreement with the nominal value of 1.1×10^{-14} [2] indicating a spectrometer background level of $<10^{-14}$. To determine the ^{10}Be background level of the spectrometer more accurately low-level ^{10}Be blank material needs to be tested.

Although the spectrometer was optimized for ^{10}Be, it should be possible to measure other radionuclides such as ^{26}Al, ^{129}I or actinides. The performance of the facility for these nuclides will be determined in the near future.

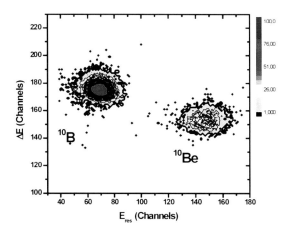

Fig. 2: *Spectrum ($\Delta E \leftrightarrow E_{res}$) of a ^{10}Be standard sample.*

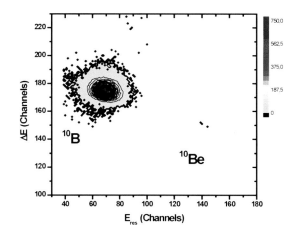

Fig. 3: *Spectrum ($\Delta E \leftrightarrow E_{res}$) of the BEBL low level standard.*

[1] C. Calcagnile et al., Nucl. Instr. & Meth. B 294 (2013) 416

[2] C. Calcagnile et al., Nucl. Instr. & Meth. B (2014) accepted

[1] *CEDAD, University of Salento, Lecce, Italy*

THE UPGRADE OF THE PKU AMS – THE NEXT STEP

Reconstruction of the detector beam line

A.M. Müller, X. Ding[1], D. Fu[1], K. Liu[1], M. Suter, H.-A. Synal, L. Zhou[2]

The successful [10]Be tests with the degrader foil method performed last year at the PKU AMS facility led to the decision to equip the system with an additional analyzing magnet. The magnet design is based on experience made with a similar magnet at the ETH 500 kV TANDY facility [1]. The overall [10]Be transmission should increase to about 6 - 7 % due to the energy and angular focusing properties. Additionally, [9]Be ions of the same energy as [10]Be will not reach the detector resulting in a [10]Be/[9]Be machine blank level of the order of 10^{-15}.

Ion optical calculations have been performed in order to find the optimal magnet configuration (radius: 350 mm) to fit into the existing system setup without interfering with the ^{14}C measurements (Fig. 1).

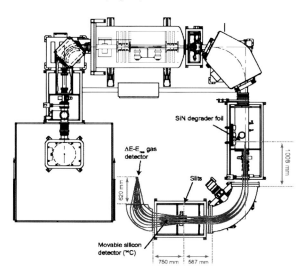

Fig. 1: *Top view of the 500 kV PKU AMS including the ion-optical calculations for the additional magnet [2].*

The conversion of the PKU AMS system was planned to be made in two steps in order to minimize the shutdown time of the facility. In June 2013, the detector beam line was reconfigured and equipped with slits in the [10]Be focal plane and a movable holder for the ^{14}C detector to be able to switch between ^{14}C and [10]Be measurement without breaking the vacuum (Fig. 2).

Fig. 2: *The new detector beam line.*

In a next step, the new magnet and the ETH built gas ionization chamber will be installed together with the necessary infrastructure (additional supports, vacuum pumps, electronics etc.), most likely in spring of 2014.

[1] A.M. Müller et al., Nucl. Instr. & Meth. B 266 (2008) 2207

[2] A.M. Müller et al., Radiocarbon 55 no. 2-3 (2013) 231

[1] *Nuclear Physics & Technology and Heavy Ion Physics, Peking University, Beijing, China*
[2] *Geography, Peking University, Beijing, China*

THE MICADAS AMS FACILITIES

Radiocarbon measurements on MICADAS in 2013

Upgrade of the myCADAS facility

Phase space measurements at myCADAS

First sample measurements with myCADAS

Rapid ^{14}C analysis by Laser Ablation AMS

Oxidize or not oxidize

Graphitization of dissolved inorganic carbon

Ionplus – A new LIP spin-off

RADIOCARBON MEASUREMENTS ON MICADAS IN 2013

Performance and sample statistics

Scientific and technical staff, Laboratory of Ion Beam Physics

At the end of 2012, the MICADAS system was equipped with a new gas handling box, to which, in 2013, a carbonate handling system was added to allow the sampling of CO_2 from the headspace of septa-sealed vials.

We also installed a stable isotope mass spectrometer for precise $\delta^{13}C$ measurements. Already, the new spectrometer produces routinely reliable $\delta^{13}C$ results. The goal for 2014 is to integrate it into our measurements procedures so that $\delta^{13}C$ and radiocarbon measurements can be performed simultaneously on the same sample of gas.

The MICADAS AMS facility operated on 330 days with 5500 hours spent on measuring samples (Fig. 2). This corresponds to 60 % of the full year being used as pure measurement time, a 20 % increase over last year! More than 6600 samples and standards (including some replicates), were measured, which was a small increase over the previous year. Gas measurements leveled out for the first time after significant increases in the years.

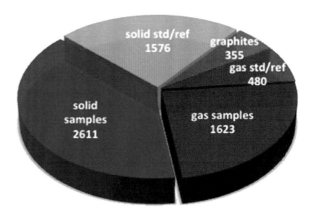

Fig. 2: Samples and standards measured on the MICADAS prepared at ETH Zurich as solid graphite samples (blue) or received as graphite targets (gray). Samples in red were measured with the gas ion source.

The slowdown in the increase of measured samples was expected as the University of Bern MICADAS AMS system came on line. Previously we had measured more than 1000 gas samples a year for them. The majority of samples this year were measured for our partners from the Department of Earth Sciences at ETH Zurich. Members and students produced about 1000 solid and 1400 gas samples.

Fig. 1: *The MICADAS ion source with the new gas interface including (1) a cracker changer for the measurement of CO_2 in glass ampoules, (2) a carbonate handling system, (3) an elemental analyzer and (4) a stable isotope mass spectrometer.*

UPGRADE OF THE MYCADAS FACILITY

Optimizing differential pumping and stripper design

M. Seiler, S. Maxeiner, H.-A. Synal

The myCADAS facility was upgraded during this year to reduce background and to improve measurement stability. An important issue was optimizing the gas flow in the stripper region, because bad vacuum conditions in the following spectrometer (electrostatic analyser, ESA) could cause additional background [1].

The pressure measured in the ESA, right after the terminal, was remarkably higher than expected from gas flow calculations. Although a simplified geometry was considered, the main reason for the discrepancy was that the differential pumping setup did not perform as intended. The inner stripper chamber is only bolted to the housing leaving gaps and holes for the stripper gas to escape through and to increase the pressure in the region leading to the spectrometers (T2). Sealing these openings with aluminium tape (Fig. 1) reduced the pressure in the ESA by a factor of 5.

Fig. 1: *Picture of inner stripper chamber with schematics of stripper and gas flow into volumes pumped by different turbo pumps (T1, T2).*

The now lower pressure in the T2 and ESA regions allowed the investigation of background due to residual gas in the ESA. A gas inlet was installed to vary the pressure in the ESA independently from the stripper density. Measured ratios of test blanks (Fig. 2) and

standards, however, showed no pressure dependence.

As higher pressure in the ESA does not seem to increase background, the reduced pressure obtained with the improved differential pumping can be sacrificed for a higher stripper gas flow associated with a larger ion beam acceptance.

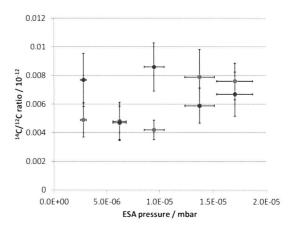

Fig. 2: *Measured $^{14}C/^{12}C$ ratios of two blank samples (red/blue) at different stripper pressure.*

A new stripper tube for a density of about 0.5 µg/cm^2, needed for molecular dissociation, was designed using molecular flow simulations (for ideal pumping) [2] as well as phase space data from the myCADAS ion source [3]. The stripper tube has a slightly increased minimal diameter of 3.2 mm, a half angle acceptance of 37 mrad, and a reduced length doubling the gas flow through the stripper. The larger acceptance of this stripper increased the transmission from about 30 % to 40 %.

[1] H.-A. Synal et al., Nucl. Instr. & Meth. B 294 (2013) 349

[2] S. Maxeiner et al., Laboratory of Ion Beam Physics Annual Report (2013) 15

[3] M. Seiler et al., Laboratory of Ion Beam Physics Annual Report (2013) 24

PHASE SPACE MEASUREMENT AT MYCADAS

Measurement of beam size for stripper optimization

M. Seiler, G. Klobe, M. Passarge, H.-A. Synal

Helium as a stripper gas allows radiocarbon measurements with reasonable efficiency at very low energy (e.g. 45 keV [1]). However, without the accelerator the beam's phase space cannot be compressed leading to additional losses in narrow ion optical elements like the stripper.

To be able to optimize the transmission through the stripper tube the beam size at the stripper location was measured by replacing the stripper with a Faraday cup (Fig. 1), which could be moved along the beam axis (z-axis).

Fig. 1: *Picture of movable Faraday cup in the stripper chamber.*

Changing the magnetic field strength of the injection magnet moves the beam horizontally across the cup opening. The varying ion current in the cup allows the calculation of the beam width at a given position z. To determine the divergence of the beam, a series of measurements at different positions z were made with the same ion source settings. With this data the phase space volume and the position of the focal point after the injection magnet could be calculated.

It turned out that in the used ion source configuration the divergence angle of the beam at the stripper position was smaller than the calculated optimum. Moving the box lens of the ion source closer to the injection magnet allowed us to increase the beam divergence with a corresponding reduction of the beam size in the stripper. Measurements with the stripper tube in place showed an increase in transmission from 30 % to 33 % due to this change.

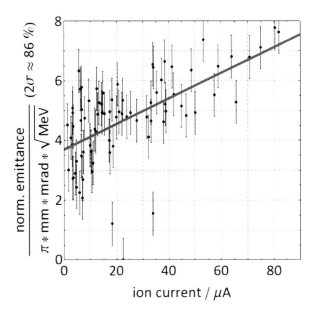

Fig. 2: *Measured emittance values of the myCADAS ion source as a function of carbon ion current.*

As expected, a dependence of the normalized emittance on the measured ion current was observed (Fig. 2). For an ion current of $\approx 40\ \mu A$, which allows performing a radiocarbon dating measurement in a reasonable time, the phase space volume is ≈ 75 mm mrad at 45 keV. This is the key parameter in the design of a stripper tube optimized for the myCADAS facility.

[1] H.-A. Synal et al., Nucl. Instr. & Meth. B 294 (2013) 349

FIRST SAMPLE MEASUREMENTS WITH THE MYCADAS

Measurement stability and reproducibility of real samples

M. Seiler, L. Wacker, H.-A. Synal

In 2013, the myCADAS was upgraded to improve background and transmission [1]. For precise measurements, however, also stability over time is important.

For this purpose, a test magazine (Tab. 1) was prepared containing 8 *NIST oxalic acid 2* samples as a reference material for normalization, 5 blank samples made of radiocarbon free phthalic anhydride, and 5 reference wood samples with an expected radiocarbon concentration of 95.7±0.2 pMC.

Label	Type	pMC	pMC err	χ^2_{red}	$\delta^{13}C$
52011.1.17	oxa2	133.71	0.29	1.28	-18.7
52011.1.18	oxa2	134.16	0.31	3.06	-15.9
52011.1.19	oxa2	134.73	0.35	1.92	-14.3
52011.1.24	oxa2	134.57	0.32	1.47	-23.8
52011.1.25	oxa2	133.38	0.33	2.22	-14.6
52011.1.26	oxa2	133.89	0.32	0.61	-18.3
52012.1.17	bl	0.53	0.02	0.84	-34.8
52012.1.19	bl	0.81	0.02	1.23	-30.3
52012.1.24	bl	0.58	0.02	0.77	-34.0
52012.1.25	bl	0.46	0.01	1.01	-29.9
52012.1.26	bl	0.47	0.02	1.18	-32.1
52796.1.1	wood	96.24	0.30	2.57	-23.2
52797.1.1	wood	94.70	0.27	4.62	-25.6
52798.1.1	wood	95.62	0.29	2.77	-20.7
52799.1.1	wood	97.61	0.30	6.68	-36.9
52800.1.1	wood	96.14	0.28	1.09	-27.0

Tab. 1: *Sample list of magazine C131010MS1 with measured $\delta^{13}C$ and pMC values.*

The extracted C⁻ ion current from the ion source was about 50 μA at the beginning of a measurement and decreased slightly over time. The average beam transmission was about 28 %. To reduce background coming from dark counts the bias voltage of the electron multiplier was lowered. This resulted in a reduced counting efficiency of about 65 % for ^{14}C.

Every sample was measured 16 times (passes) for 5 minutes each. The $^{13}C/^{12}C$ ratios showed a strong burn-in effect over the first 3 passes (Fig. 1). Although these 3 passes were not included in the data analysis [2], the $\delta^{13}C$ values (Tab. 1) show a scatter of about 0.4 % (1σ).

One sample showed a drastically lowered $\delta^{13}C$ ratio after some time resulting in a higher $^{14}C/^{12}C$ ratio after fractionation correction, a feature that was also observed in other measurements not presented here. The changes happened when the sample material was partially used up. It likely indicates a phase space dependency of the measurements leading to a variation between (identical) samples and thus limiting measurement accuracy.

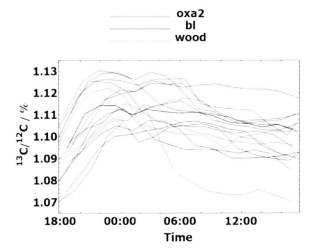

Fig. 1: *Time dependence of measured $^{13}C/^{12}C$ ratios.*

The pMC data (Tab. 1) show a larger scatter than counting statistics (including error propagated corrections such as standard normalization) would predict. The additional uncertainty of about 0.4 %, estimated with a χ^2-test, leads to an overall precision for the wood samples of about 0.6 %.

The blank level corresponds to a radiocarbon age of 40'000 to 43'000 years.

[1] M. Seiler et al., Laboratory of Ion Beam Physics Annual Report (2013) 23

[2] L. Wacker et al., Nucl. Instr. & Meth. B 268 (2010) 976

RAPID ^{14}C ANALYSIS BY LASER ABLATION AMS

Fast analysis of materials at high spatial resolution

C. Münsterer[1], L. Wacker, B. Hattendorf[1], J. Koch[1], M.Christl, R. Dietiker[1], H.-A. Synal, D. Günther[1]

By focusing a pulsed laser onto solid materials small amounts of matter can be removed and analyzed [1]. This is a rapid sampling technique which achieves a high spatial resolution in the order of 100 μm or smaller.

When applying laser ablation (LA) to carbonates a high proportion of the ablated $CaCO_3$ is converted into CO_2. The ^{14}C content of this CO_2 can be analyzed with gas ion source (GIS)-AMS [2] avoiding the time consuming and contamination prone conventional graphitization process for small radiocarbon samples. For an online ^{14}C analysis of carbonates a LA-setup (Fig. 1) was developed comprising a 193 nm ArF excimer laser, an optimized ablation cell, a designated observation unit and a gas transportation system. First measurements have been performed on marble, on pressed carbonate powder, and on stalagmite samples to assess their aptitude for LA-AMS analyses. The marble sample is considered to be a ^{14}C blank, while the pressed IAEA C2 powder has a known ^{14}C activity.

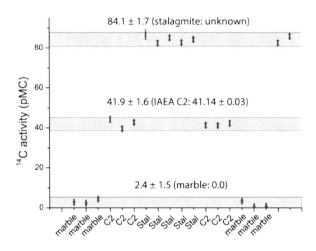

Fig. 2: *Repeatability study on several materials (grey bands: two standard deviations)*

The ^{12}C currents (HE) ranged from 0.5 to 4.8 μA. For the pressed sample currents were generally lower than for the marble and stalagmite.

The measurement sequence shown in Fig. 2 was designed to investigate possible cross-talk and to determine the measurement reproducibility. The nominal value of the IAEA C2 standard was reproduced albeit with a rather large error. The measurements of the stalagmite sample show similarly large scatter. The source of the rather large marble blank value (2.4±1.5 pmC) needs to be investigated.

The LA-AMS setup will be modified to accommodate higher ^{12}C currents to reduce measurement time and to increase counting precision.

[1]   J. Koch and D. Günther, Appl. Spectr. 65 (2011) 155A

[2]   M. Ruff et al., Radiocarbon 49 (2007) 307

^{1} D-CHAB, ETH Zurich

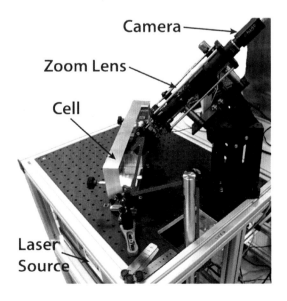

Fig. 1: *Overview of the experimental setup. The purple lines indicate the path of the laser beam.*

OXIDIZE OR NOT OXIDIZE?

Evaluation of methods for treatment of charcoal

I. Hajdas, K. Hippe, S. Ivy Ochs, M. Maurer

The radiocarbon method requires pre-preparation of the material prior to the isotopic analysis i.e. decay counting (conventional) or counting ^{14}C atoms (AMS). Since the early days of the method the standard treatment involved removal of contamination by washes in acid (dissolves carbonates), base (dissolves humic acid), and acid. The last step removes possible contamination with modern CO_2 trapped during the base step [1]. Modification of this method has been suggested, especially for old material, i.e. older than 20 ka (for review see [1]). However, replacing the ABA (Fig. 1) method with a more aggressive oxidizing method might only be necessary in some special cases, for example for material of poor preservation.

In order to assess the effectiveness of ABA we used both treatments (Fig. 1) on selected old (>20 ka) charcoal samples.

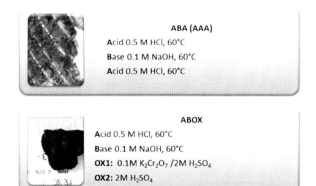

Fig. 1: *Treatment methods applied to charcoal (ABA: acid, base, acid, ABOX: acid, base, OX1, OX2).*

Samples from various archives were included in this study including charcoal from Chauvet used in an intercomparison ^{14}C dating project [2] and charred wood from 40 ka old lava flows. The main requirements were old age and sufficient material.

The results of the study are summarized in Fig. 2, which shows the offset between ^{14}C ages obtained on material treated with ABA and 2 variations of ABOX (Fig. 1 and [1])

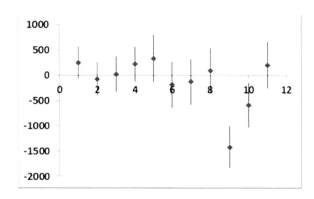

Fig. 2: *Difference (years) between ^{14}C ages of old samples (between 30 and 40 ka BP) treated with ABA and ABOX. In the case of 2 charcoal samples (lava flow) the offset was significant. A negative difference means that ABA treated samples had ages too young.*

In most cases the ABA treatment is sufficient, a finding supported by other studies [3]. However, evidence exists that some material requires an additional oxidation step [4]. This is also observed in our results (Fig. 2) where for 2 out of 12 samples the treatment with ABA resulted in ages too young. For this reason we will still continue to crosscheck very old material with duplicate preparation. It seems, however, that as long as proven otherwise there is no reason to mistrust data published in the early days, which were obtained on material treated only with ABA.

[1]   I. Hajdas, QSJ - Eiszeitalter und Gegenwart, 57 (2008) 2

[2]   A. Quiles et al., Radiocarbon (2014) in press

[3]   G.M. Santos and K. Ormsby, Radiocarbon 55 (2013) 534

[4]   M.I. Bird et al., Radiocarbon 41 (1999) 127

GRAPHITIZATION OF DISSOLVED INORGANIC CARBON

Procedures for AGE preparation of water samples for DI[14]C analysis

T. Blattmann[1], C. McIntyre, L. Wacker, T. Eglinton[1]

Dissolved inorganic carbon (DIC) encompasses dissolved carbon dioxide, carbonic acid, bicarbonate, and carbonate present in water. The ^{14}C composition of DIC (DI[14]C) is a powerful tool for assessing DIC turnover or provenance and provides a means for understanding exchanges between aquatic organisms and inorganic carbon species. A procedure for DIC sample preparation is described.

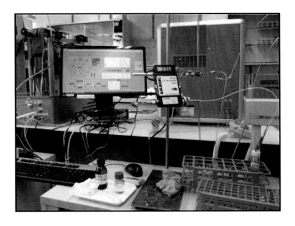

Fig. 1: *Purging and graphitization setup.*

Sample collection procedures followed after a modified WHOI-NOSAMS protocol [1]. Water collected in a Niskin bottle was discharged in small amounts into a pre-combusted (350 °C) 1 L glass bottle. This water was discarded after wetting the sides of the bottle several times. After filling the bottle with sample water to about 1 cm below the stopper height, 100 µl of saturated $HgCl_2$ solution was added to poison microbial activity to avoid post sample collection changes in DI[14]C. The bottle was sealed airtight with Apiezon N grease applied to the glass stopper. Water samples were stored at room temperature for several weeks.

40 ml aliquots of the water samples were transferred to flat-bottomed 60 ml vials with Teflon-coated silicone septa caps (Fig. 1). Needles for helium inflow and outflow were inserted into the septa, purging the headspace for 2 - 4 minutes at 100 ml/min. 1 ml of 85 % orthophosphoric acid was injected into the sample vial. Capturing of CO_2 on a zeolite trap on an AGE system took place over the course of 8 min (Fig. 2). During capture, He flow rates would drop to 96 - 98 ml/min. Afterwards, CO_2 was released to an AGE system [2] for graphitization.

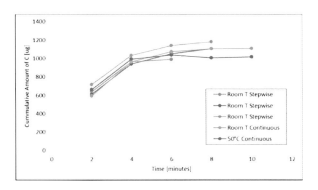

Fig. 2: *Purging experiments to determine an optimal purging time show a plateau in released carbon amount after 8 minutes. No difference was observed between stepwise, continuous, and heated experiments.*

The DIC concentrations in the water samples investigated ranged 0.04 - 0.08 mg C/ml. Samples with greatly different DIC concentrations may need to be purged under different conditions. A graphite duplicate of one water sample reproduced DI[14]C within error. Two IAEA-C1 carbonate processing blanks yielded 0.20±0.02 and 0.17±0.01 pmC.

[1]   www.whoi.edu/nosams/Sample_Prep-Inorganic carbon

[2]   L. Wacker et al., Nucl. Instr. & Meth. B 268 (2010) 931

---

[1] *Geology, ETH Zurich*

IONPLUS – A NEW LIP SPIN-OFF

A new spin-off company for ^{14}C lab instruments was founded in 2013

S. Fahrni, P. Amrein, J. Bourquin, A. Herrmann, A. Müller, R. Pfenninger M. Suter, H.-A. Synal, L. Wacker

In March 2013, Ionplus AG [1] was founded as a spin-off company of the Laboratory of Ion Beam Physics (LIP) at ETH Zurich. The young company of co-founders Martin Suter, Arnold Müller, Joël Bourquin, Lukas Wacker, Rudolf Pfenninger, Hans-Arno Synal and Andreas Herrmann (Fig. 1, from left to right) develops, builds and distributes innovative instruments in the field of radiocarbon sample preparation and measurements. With its wealth of experience and know-how in ^{14}C dating and the design of measurement systems, LIP and its members provide an excellent basis for these developments.

Fig. 1: *Co-founders of the Ionplus AG (without Andreas Herrmann).*

The products of CEO Joël Bourquin and first employees Pascal Amrein and Simon Fahrni aim at a fast and efficient processing of samples and cover virtually the entire range of dedicated ^{14}C laboratory equipment. For example, the company builds and sells automated carbonate handling systems, pneumatic sample presses and tube sealing lines, but also more complex instruments such as a gas interface system (GIS) and a fully automated graphitization system [2] (AGE, Fig. 2). Key features of these products are

a high degree of automation and versatility as well as a compact and user-friendly design achieved through excellent engineering.

Fig. 2: *The Automated Graphitization Equipment (AGE) is the first instrument to fully automatically combust and graphitize carbon samples.*

While the company's office is located at ETH Zürich, Hönggerberg for now, it will move off campus in mid to late 2015. In the new location the production will be expanded to the MIni CArbon DAting System (MICADAS) [3], the most compact ^{14}C-AMS system commercially available to date.

In its first year in business, the young company managed to sell a considerable number of instruments (including 3 AGE and 2 GIS systems), mostly within Europe but also overseas. With full order books, Ionplus is about to begin its second year in business and is hoping to meet the future needs of radiocarbon laboratories around the world.

[1] www.ionplus.com
[2] L. Wacker et al., Nucl. Instr. & Meth. B 268, 7-8 (2010) 931
[3] H.-A. Synal et al., Nucl. Instr. and Meth. B 259, 1 (2007) 7

RADIOCARBON APPLICATIONS

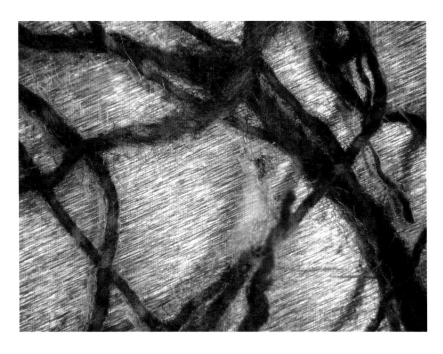

^{14}C samples in 2013

The 774/775 AD event in the southern hemisphere

Radiocarbon dating to a single year

High resolution ^{14}C records from stalagmites

^{14}C mapping of marine surface sediment

Tracing of organic carbon with carbon isotopes

Organic carbon transported by the Yellow River

On the stratigraphic integrity of leaf waxes

"True" ages of peat layers on the Isola delta

Middle Würm radiocarbon chronologies

^{14}C SAMPLES IN 2013

Overview of samples types prepared and measured at ETH

I. Hajdas, C. Biechele, G. Bonani, C. McIntyre, M. Maurer, H-A. Synal, L. Wacker

The number of samples analyzed for ^{14}C in 2013 increased again, following a trend that has been observed for previous years (Tab. 1). More than 6200 analyses were performed which included both gas and graphite samples prepared by ETH/LIP, and externally prepared graphite targets that were sent to us. Standards and blanks prepared for both the gas and graphite samples constitute 25 % of the total.

Sample Type	2010	2011	2012	2013
Standards (OXA1&2)	*290*	*441*	*615*	*740*
Blanks	*283*	*317*	*529*	*642*
IAEA	*130*	*130*	*202*	*194*
Subtotal	703	888	1346	1576
Archaeology	*764*	*669*	*808*	*504*
Past Climate	*191*	*294*	*275*	*109*
Environment	*145*	*73*	*30*	*67*
Art	*169*	*264*	*178*	*171*
Geochronology	*26*	*31*	*246*	*175*
Subtotal	1405	1589	2004	1026
LIP/ETH Projects	*110*	*258*	*467*	*1417*
Gas Samples	*993*	*1040*	*2303*	*2103*
External Targets	*753*	*949*	*307*	*83*
Grand Total	**3854**	**4466**	**5960**	**6205**

***Tab. 1**: Number of samples from 2010 to 2013 prepared and measured for various applications. ETH projects include samples that are part of research from the Department of Earth Sciences (ERDW).*

Overall, more than 4000 graphite targets were produced in 2013. The number of commercial samples decreased this year by almost 1000, however this was offset by an equivalent increase from ETH projects. Specifically, a significant increase of samples came from members and students of the Geological Institute and the Biogeosciences group therein.

A breakdown of the different types of materials submitted as commercial samples is given in Figure 1. Samples are frequently prepared in replicate or using more than one method [1]. As such, nearly 1200 targets were measured to complete the analysis.

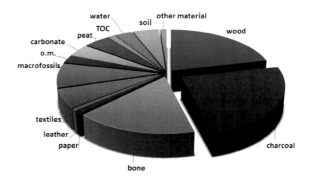

***Fig. 1**: Types of sample material submitted for preparation and ^{14}C dating in 2013. The most common are charcoal (23 %), wood (21 %) and bone (17 %). Carbonates include shells, pearls, stalagmites, foraminifera, etc. Other material includes antler, oil, wine, hair etc.*

Archaeology, climate studies and geochronology were the leading applications in 2013 (Tab. 1). Art pieces are less common but remain significant and an interesting part of our work. Projects that commenced this year were work on the dating of very old samples of wood, charcoal and bone/ivory [1, 2].

In addition to solid graphite samples, 2103 measurements were performed with gas samples.

[1] I. Hajdas et al., Laboratory of Ion Beam Physics Annual Report (2013) 27

[2] K. Hippe et al., Laboratory of Ion Beam Physics Annual Report (2013) 41

THE 774/775 AD EVENT IN THE SOUTHERN HEMISPHERE

Comparing atmospheric ^{14}C of the southern and northern hemispheres

D. Güttler, J. Beer[1], N. Bleicher[2], G. Boswijk[3], A. G. Hogg[4], J.G. Palmer[5], L. Wacker, J. Wunder[6]

Miyake et al. [1] reported in 2012 a sudden and strong increase of the atmospheric radiocarbon ^{14}C content of 1.2 % between AD 774 and 775, measured in Japanese trees. While their findings were quickly confirmed in the German oak chronology for the northern hemisphere (NH), we present here the first southern hemisphere (SH) record covering the period AD 760 - 787, measured in New Zealand Kauri wood (Agathis australis).

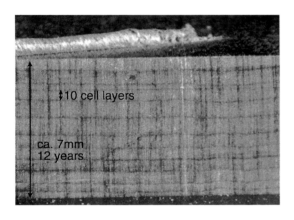

Fig. 1: *Microscope picture of the Kauri wood cross-section as recorded during separation of single tree rings by means of a microtome.*

Radiocarbon concentrations were measured on dendrochronologically dated annual tree ring samples cut with a microtome from New Zealand Kauri wood (Fig. 1). Cellulose was isolated from the tree ring samples, then converted to graphite and radiocarbon analyzed on the MICADAS [2]. Fig. 2 shows a compilation of the new ^{14}C AMS measurements on SH Kauri wood together with previous measurements on NH samples [1, 3].

The rapid change of the Δ^{14}C signal obtained for the SH is in agreement with the previous findings for the NH. The jump in Δ^{14}C within 775 AD is 14 ‰ and a total change of 19 ‰ in Δ^{14}C within 775 - 776 AD is measured. Based on the

consistency of the signals from both hemispheres it is most likely that the impact of the potential cosmic event that caused the Δ^{14}C increase affected equally both hemispheres. A short γ-ray burst in our galaxy was discussed as a possible source of the NH signal [4]. With our new data we can conclude that such an event would most probably have occurred where the ecliptic plane of our solar system cuts through the Galaxy. A large solar flare event as cause [3], however, would affect both hemispheres.

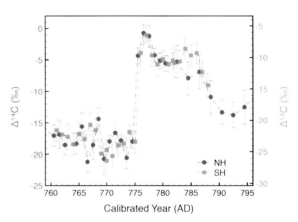

Fig. 2: *Δ^{14}C values obtained for the southern (green) and the northern hemisphere (blue).*

[1]   F. Miyake et al., Nature 486 (2012) 240
[2]   L. Wacker et al., Radiocarbon 52 (2010) 252
[3]   I.G. Usoskin et al., A&A 552 (2013) L3
[4]   V.V. Hambaryan and R. Neuhäuser, MNRAS 430 (2013) 32

[1] *Radiocative Tracers, Eawag, Dübendorf*
[2] *Laboratory for Dendrochronology, Zurich*
[3] *Environmental Science, University of Auckland, New Zealand*
[4] *Science and Engineering, University of Waikato, New Zealand*
[5] *Gondwana Tree-Ring Laboratory, Christchurch, New Zealand*
[6] *Terrestrial Ecosystems, ETH Zurich*

RADIOCARBON DATING TO A SINGLE YEAR

Dating in times of rapid atmospheric radiocarbon changes

L. Wacker, D. Güttler, J. Goll[1], J. P. Hurni[2], H.-A. Synal, N. Walti

Wooden objects such as support beams in old buildings can be radiocarbon dated with a precision generally not better than 10 calendar years even when conventional wiggle-matching onto the present IntCal radiocarbon calibration curve is applied. More precise dating is possible only with annually resolved radiocarbon calibration data, particularly in periods of rapid changes in atmospheric radiocarbon concentration. The recently observed jump in atmospheric radiocarbon concentration of 1.5 % between 774 and 775 AD, though expected to be a rare occasion, is a good example for such a rapid change [1]. We demonstrated exemplarily this possibility by radiocarbon dating a timber in the historically important and well-preserved Holy Cross chapel of the convent St. John the Baptist in Val Müstair, Switzerland (see Fig. 1).

Fig. 1: *The Holy Cross chapel of the St. John convent.*

Cellulose was isolated from 7 subsequent tree ring samples of a wall lath in the chapel, then converted to graphite and radiocarbon analyzed on the MICADAS [2].

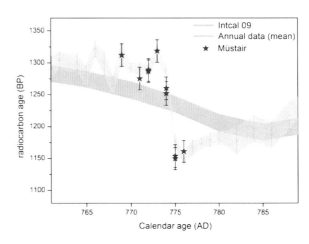

Fig. 2: *The radiocarbon ages of the tree ring samples are matched to annually resolved tree-ring data of trees of Japan and Europe.*

Our data is compared in Fig. 2 with the IntCal calibration curve and a high-temporal resolution calibration curve (HR-curve). The latter is deduced from the mean values of dendro-chronologically dated, annually resolved tree ring measurements from Japan [1] and Europe [3]. The measured radiocarbon concentrations of the timber samples are matched to the HR-curve such that the chi-square becomes minimal. A perfect match with a chi-square of 9.1 (n=10, χ^2=0.91) is obtained when the last ring (waney edge) of the wall lath is from 785 AD. A shift of one year (to 786 AD) is the second best option with an increase of χ^2 to 36.6.

The Holy Cross chapel in Müstair is herewith one of the most precisely dated Carolingian buildings.

[1]   F. Miyake et al., Nature 486 (2012) 240
[2]   L. Wacker et al., Radiocarbon 52 (2010) 252
[3]   I.G. Usoskin et al., A&A 552 (2013) L3

[1] *Archäologischer Dienst Graubünden, Müstair*
[2] *Lab. Romand de Dendrochronologie, Moudon*

HIGH-RESOLUTION ^{14}C RECORDS FROM STALAGMITES

Proxy for karst hydrology or recorder of atmospheric ^{14}C?

F.A. Lechleitner[1], C. McIntyre, T. Eglinton[1], J.U.L. Baldini[2]

Stalagmites have become one of the most promising and powerful archives for paleoclimate studies, providing multi-proxy reconstructions of climatic and environmental conditions at often unprecedented resolution and precision. Radiocarbon incorporated in stalagmite carbonate is a promising new proxy to trace hydrological conditions in karstic systems [1], and in some cases to help reconstruct atmospheric ^{14}C [2].

A 2000-year old stalagmite (YOK-I) from the Yok Balum cave in Belize, Central America, was sampled at annual-to-decadal resolution by continuously drilling, and yielding carbonate powders (Fig. 1).

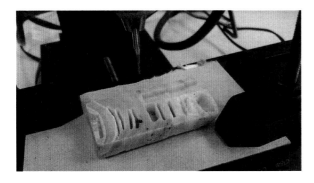

Fig. 1: *Drilling of carbonate samples from a small stalagmite slab. A semi-automatic drill is used (Sherline 5400 Deluxe).*

The resulting record shows exceptional detail in the measured ^{14}C signal, due to the small sample size required by the method and possibility of analyzing a large number of samples (Fig. 2). The interval that has been studied covers the period of the downfall of the Classic Mayan culture around 1100 AD, which has been related to a prolonged "super-drought" in Central America, documented in the stable oxygen isotope record from YOK-I [3]. The ^{14}C record is remarkably similar in shape to the stable carbon isotope record (δ^{13}C), both

showing a significant drought occurring during the period of the Classic Maya collapse (ca. 650 - 1100 AD, Fig. 2). Preliminary interpretation suggests that ^{14}C recorded in YOK-I is related to changing degrees of open vs. closed system behavior in the overlying karst, resulting in more or less equilibration with atmospheric ^{14}C [1].

At the same time, the YOK-I ^{14}C record closely resembles (with a slight lag) the atmospheric calibration curve during that period, suggesting fast signal transfer through the karst and close coupling to the surface/atmosphere.

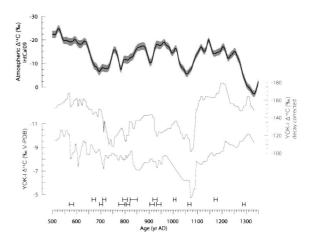

Fig. 2: *Δ^{14}C (blue, decay corrected) and δ^{13}C (red) records from YOK-I, atmospheric ^{14}C calibration curve (IntCal09) for the interval 500-1350 AD.*

[1] M.L. Griffiths et al., Quat. Geol. 14 (2012) 81

[2] D.L. Hoffmann et al., Earth Planet. Sci. Lett. 289 (2010) 1

[3] D.J. Kennett et al., Science 338 (2012) 788

[1] *Geology, ETH Zurich*
[2] *Earth Sciences, Durham University, United Kingdom*

^{14}C MAPPING OF MARINE SURFACE SEDIMENTS

Sediment resuspension as a key control on organic carbon burial

R. Bao[1], C. McIntyre[1], M. Zhao[2,3], T. Eglinton[1]

Continental shelves form a crucial interface between the land and the ocean, receiving organic carbon inputs from both reservoirs. These systems account for about 90 % of global organic carbon burial in the modern oceans [1], however considerable uncertainty remains concerning the source and fate of organic carbon delivered to, and produced over continental shelves. In particular, controls on spatial variability in the content and composition of sedimentary organic matter on continental shelves remain uncertain [2, 3]. In addition to the magnitude and nature of organic matter supply from terrestrial sources and from surface ocean productivity, there is evidence that hydrodynamic processes and physical protection mechanisms play a critical role in influencing the dispersal and eventual burial of organic matter on the continental shelf [3].

Through combined organic geochemical and sediment fabric analysis of bulk surface sediment and corresponding grain size fractions, we show that sedimentological processes in the Chinese marginal seas (CMS) exert important control on content and age of organic carbon (Fig. 1). Variations in hydrodynamic sorting and physical protection of organic matter are observed in relation to grain size and related sedimentary properties. An extensive survey of CMS surface sediments reveals that pre-aged organic carbon is associated with distinct grain sizes. Organic carbon contents and isotopic compositions coupled with grain size distributions suggest that pre-aged organic matter with relatively high organic carbon content and depleted stable carbon isotope signatures accumulates on the inner shelf and in high energy regions. This indicates that it derives from terrestrial organic carbon that is closely associated with finer grain size (e.g., 20-32 μm) particles and coarse biological detritus. As a consequence of protracted entrainment in

deposition-resuspension loops, the organic matter becomes more refractory and subject to widespread dispersal via transport within the benthic nepheloid layer. These results shed new light on the sources and fate of organic matter on continental shelf seas, and on controls on organic carbon preservation in continental shelf sediments. Pre-aged organic carbon may be an important consideration in developing budgets of global terrestrial organic carbon burial on the continental shelves.

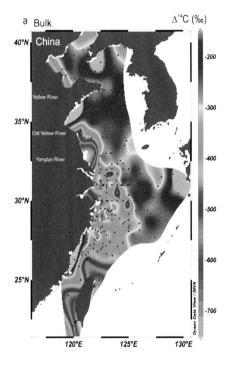

Fig. 1: *$\Delta^{14}C_{org}$ mapping of bulk surface sediments (0 -2 cm) of Chinese marginal seas.*

[1] J.I. Hedges et al., Mar. Chem. 49 (1995) 81
[2] R.G. Keil et al., Nature 370 (1994) 549
[3] J. E. Vonk et al., Nature 489 (2012) 137

[1] *Geology, ETH Zurich*
[2] *Marine Chemistry, Qingdao, China*
[3] *Center of Marine Sci. and Techn., Qingdao, China*

TRACING OF ORGANIC CARBON WITH CARBON ISOTOPES

Provenance analysis in Lake Constance with stable and radiocarbon

T. Blattmann[1], M. Wessels[2], M. Plötze[3], C. McIntyre, T. Eglinton[1]

In this study, the source-to-sink transit of organic carbon (OC) in the Lake Constance catchment and basin was investigated. Past studies have demonstrated the utility of using stable carbon isotopic compositions to shed light on terrestrial versus lacustrine OC input [1]. In this study, the addition of radiocarbon isotopic information has revealed a previously overlooked, major OC source.

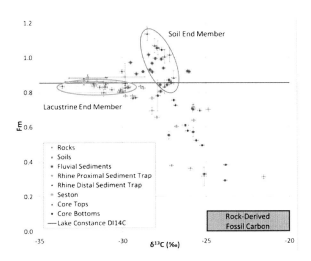

Fig. 1: *^{14}C versus ^{13}C isotopic composition of bulk OC from different settings in and around Lake Constance. Down core sediments are not decay corrected (does not change their positions appreciably). From the observations, three major OC sources become evident: (1) lacustrine end member reflecting aquatic primary productivity, (2) soil end member, and (3) a rock-derived fossil OC input from eroding shales in the Rhine catchment.*

Fluvial sediments resemble either a soil composition or lie in between a fossil and soil OC signature (Fig. 1). Seston (suspended particulate material from the water column) and Rhine-distal sediment trap material primarily show a lacustrine fingerprint. The lacustrine origin of the relatively radiocarbon depleted composition was confirmed by measurements

of dissolved inorganic carbon from the Lake Constance water column. The Rhine proximal sediment trap scatters in between the three end members. Core tops lie near the lacustrine end member except for sediments collected by the Rhine outflow, which reflect a strong terrestrial signal. The isotopic makeup of older lake sediments in the lower parts of the cores drift toward the rock and soil end members reflecting the preferential oxidation of labile lacustrine OC.

The previously overlooked fossil OC component must stem largely from the Bündnerschiefer Formation, which, due to its highly erodible nature, delivers large amounts of sediment material to the Rhine [2]. The transit of fossil OC through the Earth surface carbon cycle without its oxidation is considered an important mechanism by which atmospheric O_2 content has risen over geologic time [3].

Using coupled stable and radiocarbon isotope data, the provenance of OC cycled through sedimentary systems can be assessed. Additionally, with isotope mass balance equations, the relative contributions of the different OC sources can be quantified.

[1] N. Fuentes et al., Limnologica 43 (2013) 122

[2] A. Steudel, Schriften des Vereins für Geschichte des Bodensees und seiner Umgebung 5 (1874) 71

[3] V. Galy et al., Science 322 (2008) 943

[1] *Geology, ETH Zurich*
[2] *Institut für Seenforschung, Langenargen, Germany*
[3] *Geotechnical Engineering, ETH Zurich*

ORGANIC CARBON TRANSPORTED BY THE YELLOW RIVER

Temporal variations in the ^{14}C composition and age

S. Tao[1,2], T. Eglinton[2], D. Montluçon[2], C. McIntyre, M. Zhao[1]

The Yellow River is one of the world's most turbid major fluvial systems, delivering over 1×10^9 t of sediments annually into the Chinese marginal seas, and accounting for about 7 % of the global sediment flux to the ocean. Organic carbon (OC) carried by the Yellow River therefore may play a significant role in the global and regional organic carbon cycle. Near-surface suspended particulate organic matter (POM) samples were collected nearby the Lijing Station, upstream of the mouth of the Yellow River (Dongying) between June 2011 and June 2012 (Fig. 1).

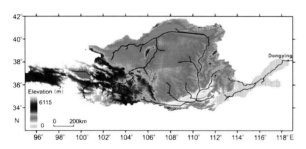

Fig. 1: *Map showing the sampling site (indicated by the red filled dot) and the drainage basin of the Yellow River.*

Particulate organic carbon (POC) in the Yellow River exhibits relatively uniform and old radiocarbon ages (4000 - 4600 ^{14}C years) (Fig. 2). Radiocarbon ages of short-chain (C_{16-18}) fatty acids (FA) were variable and relatively young, suggesting aquatic and bacterial inputs for these FA, however, the addition of higher plant FAs may be controlling the ultimate ages of these compounds in the fluvial POM pool. In contrast, lignin-derived phenols (vascular plant component) transported by the Yellow River were variable but generally younger in terms of ^{14}C age (1780 years to modern, ave. 725 years). The younger radiocarbon age of lignin phenols either reflects the contribution of woody debris or different transport dynamics. In contrast,

plant wax lipids ($C_{26+28+30}$Fatty Acids, $C_{24+26+28}$Alkanols and C_{29+31}Alkanes) are associated with fine-grained minerals and are preferentially stabilized in deep mineral soils. Their greater radiocarbon ages reflect the turn-over time of a soil carbon pool. Furthermore, substantially older radiocarbon ages were observed for long-chain even and odd n-alkanes, implying ^{14}C dilution by ancient source materials derived from sedimentary rock erosion (Fig. 2). It was possible to correct ^{14}C ages for this "petrogenic" contribution to plant-derived n-C_{29+31} Alkane based on isotopic mass balance.

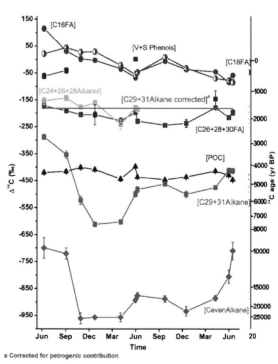

Fig. 2: *Contrasting radiocarbon contents (expressed as $\Delta^{14}C$ and conventional ^{14}C age) of various OC components compared with bulk OC.*

[1] Marine Chemistry, Ocean University of China, Qingdao, China
[2] Geology, ETH Zurich

ON THE STRATIGRAPHIC INTEGRITY OF LEAF WAXES

^{14}C dating of *n*-alkanes and *n*-carboxylic acids in loess

C. Häggi[1], R. Zech[1], C. McIntyre, M. Zech[2], T. Eglinton[1]

Paleoenvironmental reconstructions based on epicuticular leaf wax biomarkers from loess-paleosol sections are a novel tool used in paleoclimatology. The premise of such reconstruction is the synsedimentary deposition of these biomarkers and dust. To test the accuracy of this assumption, we used compound specific radiocarbon dating of long chain *n*-alkanes and *n*-carboxylic acids and compared them to the deposition age of the surrounding sediment [1].

We analyzed four samples from the Crvenka loess-paleosol section, Serbia. The samples consisted of one top soil sample (Cr1), two loess samples from the Last Glacial Maximum (LGM) (Cr10), and the marine isotope stage boundary MIS2/MIS3 (Cr20) and one from the MIS5 paleosol (Cr 40).

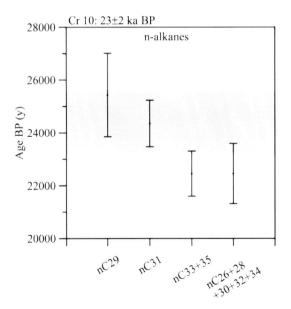

Fig. 1: *Results for the LGM sample. The grey bar indicates the sediment age [2].*

The results in Figs. 1 and 2 show that that the compound specific ages of *n*-alkanes and *n*-carboxylic acids in the loess samples are in very

good agreement with the sedimentation ages of the loess derived from optically stimulated luminescence (OSL) [2]. Cr40 approached the expected carbon dead values, with a fraction modern of only 1 %.

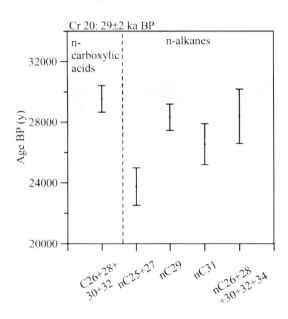

Fig. 2: *Results for sample Cr20.*

Our data thus indicates that there is little to no post sedimentary incorporation of long chain leaf waxes in the Crvenka loess profile. The incorporation of older, reworked material, that would have less of an impact on the observed radiocarbon ages, is also unlikely, since the even chain *n*-alkanes show similar ages to the dominant odd chain homologues.

[1]   Ch. Häggi et al., Biogeosci. Discussion 10 (2013) 16903

[2]   T. Stevens et al., Quat. Sci. Rev. 30 (2011) 662

---

[1] *Geology, ETH Zurich*
[2] *Geomorphology, University of Bayreuth, Germany*

"TRUE" AGES OF PEAT LAYERS ON THE ISOLA DELTA

Radiocarbon age dating of peat layers and paleosols in sediment cores

R. Grischott[1], F. Kober[2], K. Hippe, I. Hajdas, S. Ivy-Ochs, S. Willett[1]

Previously radiocarbon dated sediment cores from the subaerial delta Isola in the Upper Engadine, Switzerland (Fig. 1), revealed several age inversions [1]. These inconsistencies were based on ages from bulk peat and wood sampled from peat and paleosol layers suggesting quiet phases alternated with sandy gravels which are interpreted as fluvial sediments.

Fig. 1: *Lake Sils with the subaerial Isola delta and the small village Isola at its apex (up-front). The first age model was established for Core 5.*

We re-sampled the peat layers in newly acquired cores and obtained for the same sediment depth inconsistent ^{14}C ages on bulk peat and wood, with peat being too old (Fig. 2). Field observations revealed outcrops with peat and organic matter in the upper catchment of the Fedoz River. By eroding these peat deposits, organic material with a presumably lower ^{14}C activity is remobilized and then partly deposited on the delta. Incorporation of this reworked material yields too old ages for the Isola delta peat samples.

In order to avoid incorporation of reworked material, distinct terrestrial macrofossils such as seeds, fruits and needles (e.g. larch) were sampled. The timberline sets a natural upper limit for abundance of seeds, fruits and needles released and transported with the river, therefore long temporary storage of such

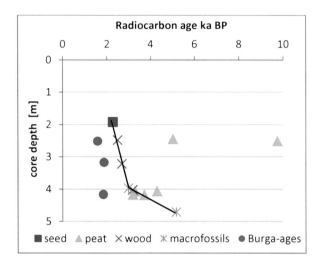

Fig. 2: *Age-depth model of core 5 based on repeated sampling of peat layers and paleosoils. Burga-ages reported after [1].*

samples in the catchment is rather unlikely. Consequently, ^{14}C activities of these samples are considered to show "true" content and ages (indicated with the dark line in Fig. 2) and thus allow construction of a reliable age depth model.

In the context of the delta evolution, we can allocate a short sedimentation time to the fluvial sediments deposited between the dated layers. This means deposition was fast, likely during floods. The large time gap between the two lowermost samples can be explained by ongoing erosion and deposition of fluvial sediments, e.g. by the presence of a stationary active channel at the core location.

[1]   C.A. Burga et al., Schlussbericht NFP 31, vdf-verlag (1997)

---

[1] *Geology, ETH Zurich*
[2] *NAGRA, Wettingen*

MIDDLE WÜRM RADIOCARBON CHRONOLOGIES

First results from the TiMIS Project

K. Hippe, I. Hajdas, S. Ivy-Ochs, M. Maisch[1]

The COST (EU project) funded TiMIS project aims at refining the radiocarbon chronology of the middle part of the last glacial cycle (middle Würm: 50 to 25 ka ago) focussing on the records preserved in the Swiss Alpine forelands where huge piedmont glaciers expanded during the Last Glacial Maximum (LGM). We perform radiocarbon dating of selected key sites in the Alpine foreland (Fig. 1) to add high-resolution chronological information crucial for understanding the phase of ice build-up just prior to the LGM and to provide important links to other records of Marine Isotope Stage 3 (MIS 3). TiMIS results are integrated into the COST Action ES0907 INITMATE.

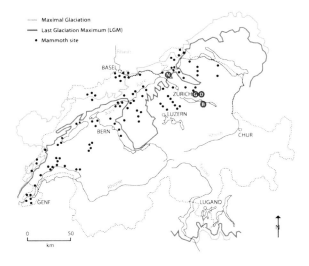

Fig. 1: Study area in the northern Alpine foreland of Switzerland, after [1]. Red circles mark the Niederweningen peat and mammoth site and the peat deposits at Gossau, Dürnten, and Bachtellen.

The Swiss foreland peat deposits are one of the best records available for the ca. 25 ka just before the LGM. Through sparse radiocarbon dating performed in the 1970s and 1980s at several compressed peat sites we know that many of these peat layers formed during MIS 3.

Preliminary ages from peat and wood samples from Dürnten (Fig. 2) agree very well with ^{14}C ages reported from the Gossau and Niederweningen sites [2, 3]. In contrast to Niederweningen, Gossau and Dürnten are both located within the LGM extent of the Linth/Rhein glacier and clearly document a phase of moderate climate and glacier absence during the early middle Würm.

Fig. 2: Sampling site of compressed peat and wood at Dürnten with radiocarbon ages.

Dating of peat is augmented by ages from archived bones and new samples. Additionally, we will compare chronologies from the northern and southern Alpine foreland by dating peat sections from drilling cores from Italy and Switzerland.

[1] H. Furrer et al., Quat. Int. 164-165 (2007) 85

[2] C. Schlüchter, Vierteljahresschrift Naturf. Ges. Zürich 132 (1987) 135

[3] I. Hajdas et al., Quat. Int. 164-165 (2007) 98

[1] *Geography, University of Zurich*

METEORIC COSMOGENIC NUCLIDES

Picture: http://www.ruhr-uni-bochum.de/sediment/pictures/paleovan_big.jpg

The Sun, our variable star

The Laschamp event at Lake Van

Authigenic Be as a tool to date clastic sediments

Meteoric ^{10}Be/^{9}Be ratios in the Amazon

Determining erosion rates with meteoric ^{10}Be

Erosion rates using meteoric ^{10}Be and $^{239+240}$Pu

THE SUN, OUR VARIABLE STAR

^{10}Be in polar ice reveals some basic properties of solar activity

J.Beer[1], F. Steinhilber[1], J. Abreu[1], U. Heikkilä[1], M. Christl, P.W. Kubik

The Sun is by far our most important source of energy. At the same time, the Sun is a variable star showing cycles such as the 11-year sunspot cycle.

To what extent does this variability affect life on Earth? In the 2009 Laboratory of Ion Beam Physics report we addressed this question for the last 4 centuries. Now, we expand this period to almost 10 millennia [1].

Solar variability is mainly caused by magneto-hydrodynamic processes within the Sun which have various consequences. Among those are the following.

(1) The magnetic activity affects the emission of electromagnetic radiation from the solar disc and herewith the total solar irradiance (TSI), the total amount of radiative energy reaching the Earth per square centimeter and second.

(2) Plasma (solar wind) streaming away from the solar surface carries magnetic fields which deflect the cosmic rays coming from interstellar space. Less cosmogenic radionuclides such as ^{10}Be and ^{14}C are produced during periods of high solar activity and vice versa. This modulation process provides the key to reconstruct the solar activity far back beyond the times of direct sunspots observations.

^{10}Be stored in ice cores provides a record of solar activity. Fig. 1 shows the solar activity derived from ^{10}Be data from Greenland. The effect of the geomagnetic field has been removed as well as all variability on time scales shorter than 100 years (blue line) and 500 years (red line).

These data reveal some interesting features:

- Practically everything we know refers only to a "very active" Sun.

- The sharp negative spikes in Fig. 1 reflect periods of very low solar activity, so-called grand minima which last between 60 and 110 years and are not equally distributed in time, but often form clusters of 2-3 events.

- Spectral analysis reveals the occurrence of multi-decadal to centennial cycles with well-defined periodicities (87, 104, 150, 208, 506 years), but highly variable amplitudes.

- There are periods of extended low or high activity (see red curve). Periods of low activity tend to coincide with colder periods (e.g. Little Ice Age) while periods of high activity correlate with warmer climatic conditions (e.g. Roman period).

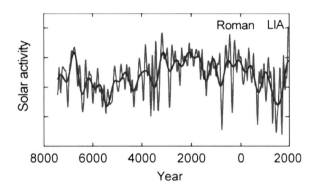

Fig. 1: *10'000 years of solar variability [1].*

Although these findings could suggest a clear relationship between solar variability and climate change it is premature to draw final conclusions because both solar forcing and climate response are not yet well enough understood. In addition, the climate system is affected by other forcings (green house, volcanic) and unforced internal variability.

[1] F. Steinhilber et al., Proc. Nat. Acad. Sci. of the USA 109 (2012) 5967

---

[1] *Radioactive Tracers, Eawag, Dübendorf*

THE LASCHAMP EVENT AT LAKE VAN

Detection of the ^{10}Be signature in lacustrine sediments

J. Lachner[1], J. Beer[1], M. Christl, M. Stockhecke[2]

The Lake Van is a saline terminal lake situated in Eastern Turkey (Fig. 1) and was sampled in the framework of the PALEOVAN ICDP project in 2010. The project aims at studying past environmental and climatic processes in this region. In parts of the core, which reaches back to ca. 600 ka [2], the sediment record is annually laminated (e.g. varved). ^{10}Be can help to constrain points in time for the age model and test the cores for their applicability as an archive resolving geomagnetic or solar activity. For this reason, ^{10}Be was studied in the time range of the Laschamp event, the geomagnetic excursion 40.7 ka ago [3].

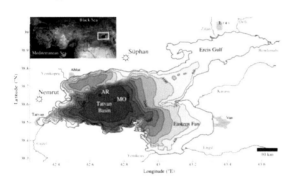

Fig. 1: *Setting of the sampling site at Ahlat Ridge (AR), Lake Van [1].*

Samples were chosen in the expected depth range for the Laschamp event. For each AMS target, 300 mg of material was prepared by adding a ^{9}Be spike, leaching with aqua regia, and purifying in several precipitation steps. The presence of ^{10}Be is predominantly related to the detrital fraction. This was confirmed by testing the chemical processing: sensitive leaching was carried out for three samples to separate ^{10}Be from the autochthonous $CaCO_3$ fraction and from the detrital fraction, and to determine the typical ^{10}Be concentration in the lacustrine $CaCO_3$. A correction dependent on the sediment's $CaCO_3$ percentage was made to estimate the detrital ^{10}Be concentration (Fig. 2).

There is no evidence for a dependency of the ^{10}Be concentrations on the lithology of the sediment. ^{10}Be concentration variations thus can be related to changes in the production.

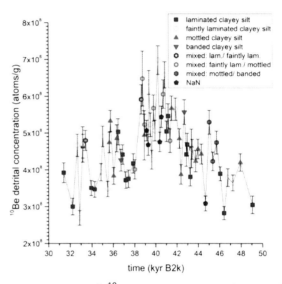

Fig. 2: *Detrital ^{10}Be concentration during the Laschamp event, sorted for the lithology of the sediment sample.*

The increase in the detrital concentration at the expected time proves the general practicality of ^{10}Be as time marker and tracer of geomagnetic activity. The high resolution of the archive promises the possibility to apply this technique to less studied excursions or solar events.

[1] M. Stockhecke et al., Palaeogeogr. Palaeoclim. Palaeoecol. 333 (2012) 148

[2] M. Stockhecke et al., Chem. Geol. (2014) in review

[3] B. Singer et al., Earth Planet. Sci. Lett. 286 (2009) 80

[1] *Radioactive Tracers, Eawag, Dübendorf*
[2] *Sedimentology, Eawag, Dübendorf*

AUTHIGENIC Be AS A TOOL TO DATE CLASTIC SEDIMENTS

Dating of late Miocene European hominids

M. Schaller[1], M. Böhme[1], J. Lachner, M. Christl, C. Maden[2]

A hominid tooth found in a sediment layer near Azmaka (Bulgaria) has been dated biochronologically to $\approx$ 7 Ma [1]. This age is younger than expected and indicates that hominids have become extinct in Europe later than previously thought.

In this study, the ages of the Miocene clastic sediment layers were determined with authigenic $^{10}Be/^{9}Be$ in order to calculate the timing for the deposition of the hominid tooth (Fig. 1). Active river sediments provide the initial starting value of the geochronologic clock. Closed-system behavior of the authigenic minerals in the sediment layer and active river sediments is required.

The clastic sediment layer associated with the hominid tooth has an authigenic $^{10}Be/^{9}Be$ age of 6.20±0.41 Ma (Fig. 2). This age agrees with the young age of the hominid tooth based on biostratigraphic dating. Therefore, if the assumptions of closed-system behavior and active river samples as the initial starting point are correct, Miocene clastic sediments seem to be datable with the method of $^{10}Be/^{9}Be$ ratios in authigenic minerals.

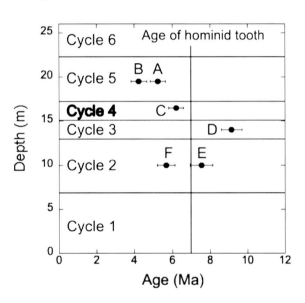

Fig. 2: *Authigenic $^{10}Be/^{9}Be$ ages of terrace samples collected from different cycles observed in the stratigraphy of the Azmaka Quarry. Cycle 4, in which the hominid tooth ($\approx$ 7 Ma, [1]) was found, has an authigenic deposition age of 6.20±0.41 Ma.*

[1] N. Spassov et al., J. Human Evolution 62 (2012) 138

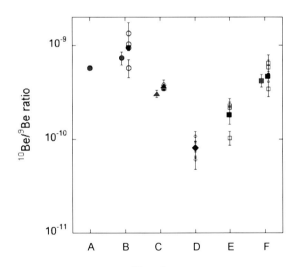

Fig. 1: *Authigenic $^{10}Be/^{9}Be$ ratios of terrace samples from different cycles (Fig. 2) of the Ahmatovo formation. The ratios were measured for combined (red symbols) and individual samples (open symbols). The combined samples are a mixture of three samples from the same depth. The averages of the individual samples (full black symbols) are in agreement with the measured values of the combined samples.*

[1] *Geology, University of Tübingen, Germany*
[2] *Geochemistry and Petrology, ETH Zurich*

METEORIC ^{10}Be/^{9}Be RATIOS IN THE AMAZON

What controls steady state between reactive and dissolved phases?

H. Wittmann[1], N. Dannhaus[1], F. v. Blanckenburg[1], J.L. Guyot[2], L. Maurice[3], N. Filizola[4], M. Christl

The ratio of the meteoric cosmogenic nuclide ^{10}Be to stable ^{9}Be has been established as a weathering and erosion proxy [1]. Meteoric ^{10}Be/^{9}Be ratios were measured in reactive phases of secondary weathering products leached from detrital Amazonian river sediment [1]. Denudation rates derived from these ^{10}Be/^{9}Be ratios are similar to those assessed by *in situ*-produced ^{10}Be from quartz [2], where source-area meteoric denudation rates agree slightly better with *in situ*-derived rates than those measured in the floodplain (see Fig. 1).

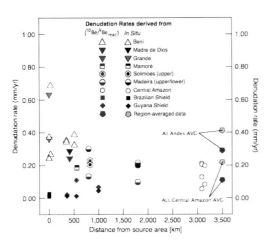

Fig. 1: *Denudation rates in the Amazon basin from reactive ^{10}Be/^{9}Be ratios (red) compared to those from in situ ^{10}Be (black) [2].*

In order to understand this new meteoric system in light of processes occurring during detrital fluvial transport and storage in lowland floodplains, we have measured corresponding dissolved ^{10}Be/^{9}Be ratios and compared them to data measured by Brown et al. [3] (Fig. 2). This comparison shows that for large rivers like the Amazon and Orinoco, reactive ^{10}Be/^{9}Be vs. dissolved ^{10}Be/^{9}Be show very good agreement.

For smaller tributaries like the Apure, La Tigra, Beni and Madre de Dios, (^{10}Be/^{9}Be)$_{reac}$ are 2 - 3 times smaller than (^{10}Be/^{9}Be)$_{diss}$. The pH values

are similar for all these rivers; thus, in smaller rivers, reactive and dissolved phases might not be fully equilibrated, i.e. the atmospheric flux may not be balanced by output. In large fluvial systems, thorough mixing of sediment and water between the channel and the floodplain may result in equilibrated phases. Resulting denudation rates seem to be robust as shown by the first-order agreement between meteoric vs. *in situ* denudation rates.

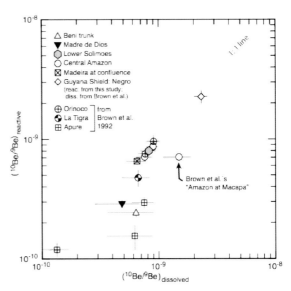

Fig. 2: ^{10}Be/^{9}Be *ratios leached from sediment ("reactive") and measured in river water ("dissolved") for Amazonian and Orinoco rivers.*

[1] F. von Blanckenburg et al., Earth Planet. Sci. Lett. 351 (2012) 295

[2] H. Wittmann et al., GSA Bull. 123 (2011) 934

[3] E.T. Brown et al., Geochim. Cosmochim. Acta 56 (1992) 1607

[1] *Earth Surface Geochem., GFZ Potsdam, Germany*
[2] *IRD, Brasilia, Brazil*
[3] *UPS (SVT-OMP), LMTG, Toulouse, France*
[4] *Geography, Fed. Univ. of Amazonas, Manaus, Brazil*

DETERMINING EROSION RATES WITH METEORIC ^{10}Be

Applying meteoric ^{10}Be to small catchments that differ in lithology

N. Dannhaus[1], F. von Blanckenburg[1], H. Wittmann[1], P. Kram[2], M. Christl

Unlike cosmogenic ^{10}Be produced *in situ*, the meteoric variety of ^{10}Be offers the possibility to measure ages and erosion rates in non-quartz bearing materials. We studied three small catchments in the Slavkov Forest, Czech Republic, that have been monitored for discharge and stream water concentrations for the last two decades [1]. The catchments are characterized by different lithologies and thus by high variation in their hydrogeochemical conditions (e.g. stream water pH). The Lysina (LYS) catchment is underlain by granite and hence has acidic output (pH ≈ 4.2). In addition, this catchment was affected by acid rain in the second half of the 20th century. The Na Zeleném (NAZ) and Pluhův Bor (PLB) catchments are dominated by mafic and ultramafic rocks, respectively, featuring neutral to alkaline environments (pH ≈6.9 at NAZ and ≈7.6 at PLB).

In these catchments we measured $[^{10}\text{Be}]_{\text{reac}}$ (the concentration of ^{10}Be adsorbed or built into secondary minerals) leached from bedload sediment to quantify catchment-wide erosion rates. Resulting reactive meteoric ^{10}Be concentrations are about 190×10^6 atoms/g (LYS) and 410×10^6 atoms/g (NAZ and PLB). We calculated erosion rates (E) from $[^{10}\text{Be}]_{\text{reac}}$ using an atmospheric ^{10}Be flux $F(^{10}\text{Be}_{\text{met}})$ of 1.05×10^6 atoms/(cm^2 a) (Q is the water discharge and K_d the solid-water partition coefficient for Be):

$$E = \frac{F\left(^{10}Be_{met}\right)}{\left[^{10}Be\right]_{reac}} - \frac{Q}{K_d}$$

In the case of Lysina, the erosion rate is corrected for the loss of beryllium into the dissolved phase due to a low pH value by using a modelled prior-acid rain pH value of 5.5 [2]. The erosion rate is highest in the granitic catchment (51 t/(km^2 a) at LYS) whereas the mafic catchments erode slower (23 t/(km^2 a) at NAZ and 25 t/(km^2 a) at PLB, Fig. 1).

As an independent control we also determined denudation rates with *in situ*-produced ^{10}Be and calculated chemical weathering fluxes using the monitoring discharge data [1] (Fig. 1).

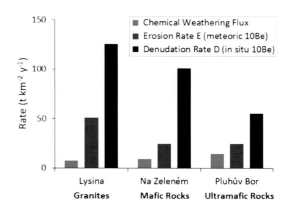

Fig. 1: *Meteoric ^{10}Be-derived erosion rates (red), chemical weathering fluxes (blue) and* in situ*-produced ^{10}Be denudation rates (t/(km^2 a)).*

Denudation rates calculated from the sum of meteoric derived erosion rates and the chemical weathering fluxes are lower than those from *in situ*-produced ^{10}Be. However, those denudation rates are in the range of *in situ*-derived denudation rates determined for catchments in middle Europe (about 50 - 180 t/(km^2 a), [3]). The observed difference in the rates could be due to different penetration depths of meteoric ^{10}Be and hence integration time scales associated with changes in erosion rate or an unrepresentative ^{10}Be flux used.

[1] P. Kram et al., Appl. Geochem. 27 (2012) 1854

[2] J. Hruška and P. Kram, HESS 7 (2003) 4

[3] M. Schaller et al., Earth Planet. Sci. Lett. 188 (2001) 441

---

[1] *Earth Surface Geochem., GFZ Potsdam, Germany*
[2] *Czech Geological Survey, Prague, Czech Republic*

EROSION RATES USING METEORIC ^{10}Be AND $^{239+240}$Pu

Alpine soils under permafrost and non-permafrost conditions

B. Zollinger[1], C. Alewell[2], K. Meusburger[2], D. Brandovà[1], P.W. Kubik, M. Ketterer[3], M. Egli[1]

Permafrost decline due to global warming might influence soil erosion processes. Erosion assessment using radionuclides can provide information on past and ongoing, i.e. time-split, processes. The focus of this work was to find out, if permafrost soils in the Swiss Alps differ in their long- (millennia: meteoric ^{10}Be) and medium-term (decades: $^{239+240}$Pu) erosion rates from non-permafrost soils and if rates have accelerated during the last few decades.

Erosion processes were estimated in permafrost soils and nearby unfrozen soils in the Alpine (sites at 2700 m asl, alpine tundra) and the sub-Alpine (sites 1800 m asl, natural forest) range of the Swiss Alps (Upper Engadine).

The ^{10}Be abundance (AMS) in the soil was estimated assuming that meteoric ^{10}Be is deposited with precipitation. Soil erosion can be estimated by comparing the effective abundance of ^{10}Be measured in the soil with the theoretically necessary abundance for the expected age [1] using equation:

$$t_{corr} = -\frac{1}{\lambda} \ln\left(1 - \lambda \frac{N}{q - \rho E m}\right)$$

m = concentration of ^{10}Be (atoms/g) in top eroding horizons, ρ = bulk density (g/cm^3), t_{corr} = expected age, E = constant erosion rate, λ = decay constant of ^{10}Be (4.997 × 10^{-7}/a), N = ^{10}Be inventory in the soil (atoms/cm^2) and q = annual deposition rate of ^{10}Be (atoms/cm^2/a). The theoretical abundance of ^{10}Be was calculated for an expected soil age of 11 ka for the Alpine "Bever" sites, 8 ka for the Alpine "Albula" sites and 16 ka for the sub-Alpine sites, respectively [2].

The evaluation of medium-term soil erosion rates is based on the differences of $^{239+240}$Pu inventories (from the 1950s and 1960s nuclear weapons tests) measured (ICP-MS) at a study site to those present at an adjacent reference site not affected by soil redistribution processes. Erosion rates were calculated [3] with:

$$L = -\left(\frac{1}{\alpha}\right)\ln\left(1 - \frac{I_{loss}}{I_{Ref}}\right)$$

L = loss of soil, $I_{loss} = I_{ref}$-I, I_{ref} = the local reference inventory as mean of all reference sites (Bq/m^2) and I = measured total inventory at the sampling point (Bq/m^2), α = obtained from a least squared exponential fit of the Pu depth profile

Site	Long-term: ^{10}Be erosion (t/km^2/a)	Medium-term: $^{239+240}$Pu erosion (t/km^2/a)
PF$_{Alpine-Bever}$	-44	-769 to +322
Non-PF$_{Alpine-Bever}$	3	20 to 486
PF$_{Alpine-Albula}$	4	-
Non-PF$_{Alpine-Albula}$	-17	-
PF$_{Subalpine}$	-45	-
Non-PF$_{Subalpine}$	-49	-

Tab. 1: *Redistribution rates from meteoric ^{10}Be (negative values indicate erosion) for permafrost (PF) and non-permafrost (non-PF) soils.*

The obtained long- and medium-term soil redistribution rates were generally low (Tab. 1). Long-term rates did not distinctly differ between the permafrost and non-permafrost sites. It however seems that soil redistribution rates have increased during the last few decades. A time-split approach seems to be promising to decipher accelerated soil erosion/accumulation caused by climate change. It seems to be likely that the higher medium-term soil redistribution rates of the last decades are the result of the ongoing climate warming. The role of methodological inherences is, however, not yet fully clarified.

[1] M. Egli et al., Geomorph. 119 (2010) 62
[2] J. Suter, Ph.D. thesis (1981) Univ. of Zurich
[3] R. Lal et al., Nucl. Instr. & Meth. B 294 (2013) 577

[1] *Geography, University of Zurich*
[2] *Environmental Geosciences, University of Basel*
[3] *Chemistry, University of Denver, USA*

"INSITU" COSMOGENIC NUCLIDES

COSMOGENIC ^{10}Be PRODUCTION RATE

Calibration from the AD 1717 rock avalanche in the Alps

N. Akçar[1], S. Ivy-Ochs, P. Deline[2], V. Alfimov, P.W. Kubik, M. Christl, C. Schlüchter[1]

The rock avalanche deposits in the upper Ferret Valley (Mont Blanc Massif, Italy), which occurred on September 12, 1717 AD, portray a high potential for calibration [1]. In this study, we calibrated the in-situ production rate based on the cosmogenic ^{10}Be from nine boulders from the rock avalanche deposits (Fig. 1).

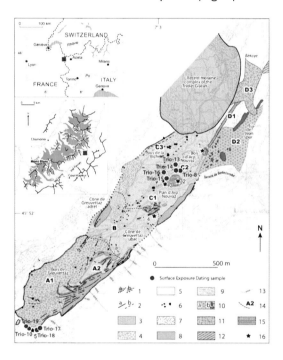

Fig. 1: *Map of the upper Ferret Valley deposits and the location of the sampled boulders for the production rate calibration. The light-brown colored area represents the extent of AD 1717 avalanche deposits, updated from [1, 2].*

Measured ^{10}Be concentrations from our samples are less than 10,000 at/g, varying from 4830±450 at/g to 5920±480 at/g. The ^{10}Be concentrations yielded a mean of 5330 at/g with a 1 σ uncertainty of 410 at/g. The calculated local production rates for each sample are between 18.4±1.8 at/(g a) and 22.4±1.8 at/(g a). We then determined the reference sea-level high latitude (SLHL) production rate for various

scaling schemes [3]: St (4.60±0.38 at/(g a)); De (4.92±0.40 at/(g a)); Du (4.88±0.40 at/(g a)); Li (5.26±0.43 at/(g a)); and Lm (4.64±0.38 at/(g a)). Although they correlate well with the global values, our production rates are clearly higher than the values obtained from other recent calibration sites (Fig. 2). This difference could be caused by a surface exposure of around 20 years at the source of the rock avalanche or by a deep muon production at depth for around 1000 years.

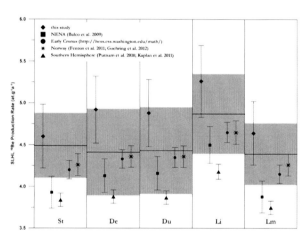

Fig. 2: *Plot of the SLHL production rates from the upper Ferret Valley and the global values from the global data set [3] according to the different scaling schemes in the online calculator. Shaded area shows the global data set.*

[1] N. Akçar et al., J. Quat. Sci. 27 (2012) 383

[2] P. Deline and M. Kirkbride, Geomorph. 103 (2009) 80

[3] G. Balco et al., Quat. Geochron. 3 (2008) 174

[1] *Geology, University of Bern*
[2] *EDYTM, University of Savoie, Le Bourget du Lac, France*

REDATING MORAINES IN THE KROMER VALLEY (AUSTRIA)

Surface exposure dating of boulders with ^{10}Be

A. Moran[1], S. Ivy-Ochs, H. Kerschner[1]

In the Kromer valley (Silvretta Mountains, western Austria), prominent moraines indicate a marked glacier advance. It was clearly larger than the Little Ice Age advance, but significantly smaller than the extent of the innermost Younger Dryas ("Egesen-Stadial") advance.

About a decade ago, several boulders on the moraine were dated with ^{10}Be to approximately 8.4 ka [1]. It was concluded that the "Kromer advance" represented the glacier reaction to the 8.2 ka event in the North Atlantic region. There was, however, conflicting evidence from other Alpine sites. A glacier advance of comparable extent in the vicinity was clearly about two thousand years older (Kartell advance [2]), while tree ring data indicated fairly small glacier extents in the Eastern Alps around 8.2 ka [3, 4].

To solve the puzzle, we decided to re-evaluate the already existing ages and to date additional boulders from further locations along the Kromer moraine system.

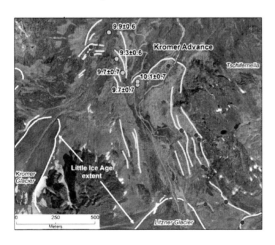

Fig. 1: *Kromer site. The green points represent the original five samples; the red points represent the new samples. Ortho images source: Federal State of Vorarlberg VOGIS.*

In total, 12 samples were collected from clast-supported boulders at the frontal moraine and the right hand lateral moraine. Although large boulders are ubiquitous at the site, sampling turned out to be extremely time consuming due to a lack of suitable quartz veins.

sample	Exposure age [ka]
KR-1	9.7±0.7
KR-2	10.1±0.7
KR-3	9.7±0.7
KR-4	9.3±0.6
KR-5	9.9±0.6

Tab. 1: *Recalculated ^{10}Be ages (ka).*

The published dates [1] were recalculated with the Northeast North America ^{10}Be production rates [4]. Recalculation yielded older ages, between 9.2 and 10.1 ka, for the five previously published dates (Tab. 1). As a consequence, the 8.2 ka event can be clearly excluded as the trigger for the glacier advance in the Kromer valley. Results from the 12 new samples will help to further constrain the timing of this early Holocene glacier advance. These results suggest that the Kartell and Kromer advances at the individual sites were not coeval and likely cannot be considered to define a single stadial.

[1]   H. Kerschner et al., Holocene 16 (2006) 7
[2]   S. Ivy-Ochs et al., Quat. Sci. Rev. 28 (2009) 2137
[3]   U.E. Joerin et al., Quat. Sci. Rev. 27 (2008) 337
[4]   K. Nicolussi and C. Schlüchter, Geology 40 (2012) 819
[5]   G. Balco et al., Quat. Geochr. 4 (2009) 93

[1] *Geography, University of Innsbruck, Austria*

DEGLACIATION HISTORY OF OBERHASLITAL

Ice flow reconstruction and ^{10}Be surface exposure dating

C. Wirsig, S. Ivy-Ochs, N. Akcar[1], J. Zasadni[2], M. Christl, C. Schlüchter[1]

Impressive, kilometer-high granite walls flank both sides of the U-shaped valley of Oberhaslital. At places, preserved polish on the smooth surfaces demonstrates that the more than 1000 m of ice that filled the valley during the Last Glacial Maximum (LGM) substantially contributed in eroding the local bedrock.

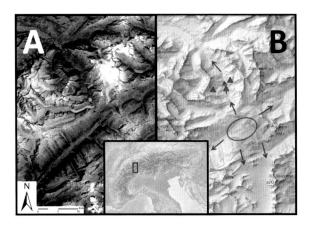

Fig. 1: *Inset: location of study site in the Alps (©Wikipedia), A: sample sites marked on satellite image centered on Grimsel Pass (©Swisstopo), B: ice flow direction marked on LGM ice surface reconstruction of the same area as A [1].*

Our investigation here combines insights of several methodological approaches. By observing trimlines and measuring the directions of glacial striations and other paleo ice-flow indicators, we aim to reconstruct extent and dynamics of glacial systems in the past. We identify two distinct generations of striations. One is dominated by an ice stream following the main valley northwards. We attribute it to the LGM glacial system connected to the Rhône glacier by ice transfluence over Grimsel Pass (Fig. 1).

A second set of striations cross-cuts the marks of the LGM. In Oberhaslital it likewise points the same direction as the main valley. Ice coming from the side valleys, however, flowed perpendicularly into Oberhaslital. During this glacial re-advance (Egesen?), the ice surface in the main valley therefore must have been significantly lower than the elevation of the hanging side valley.

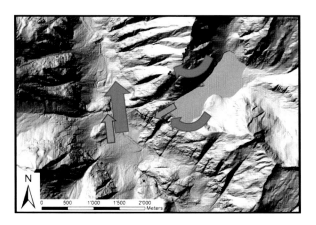

Fig. 2: *Hillshade model of the Gelmersee study area (©Swisstopo). Paleo ice-flow directions are tentatively attributed to LGM (blue) and Egesen (red) glacial systems.*

In addition, ^{10}Be surface exposure dating will provide valuable chronological information. Around Gelmersee (Fig. 2), bedrock and boulder samples were collected to address the timing of the beginning phase of ice decay after the LGM. Analyses of additional samples from Gelmersee, Gruebengletscher and Oberhaslital will serve to deduce the extent of different glacial re-advances during the Lateglacial.

[1]   A. Bini et al., Bundesamt für Landestopografie Swisstopo (2009) ISBN 978-3-302-40049-5

[1] *Geology, University of Bern*
[2] *Geology, Akademia Górniczo-Hutnicza, Krakow, Poland*

THE LATEGLACIAL AND HOLOCENE IN VAL TUOI (CH)

Geomorphological mapping, dating and glacier reconstruction

M. Messerli[1], M. Maisch[1], S. Ivy-Ochs, C. Wirsig, M. Christl

The study of glacial and periglacial landforms provides important information about the climate history of the Engadin region. Since the Last Glacial Maximum, glaciers in the Alps have advanced and retreated several times leaving traces of their activities in the field. The Val Tuoi, Grisons, Switzerland offers a wide range of glacial and periglacial landforms.

In order to reconstruct the landscape evolution and the regional climate history during the Lateglacial and the Holocene a detailed understanding of the geomorphological settings and the corresponding dating methods are compulsory. Based on current and former topographic maps, vector data, orthophotos, digital elevation models and fieldwork, a detailed geomorphological map (1:25'000) was produced and digitalized with ArcGIS. Geochronological investigation with Schmidt-hammer was applied on debris cones, moraines relictic and inactive rock glaciers.

Fig. 1: *Boulder in Val Tuoi dated to constrain the deposition age of a lateral moraine.*

To obtain absolute ages of the rock surface, samples on moraines and other deposits have been collected and dated with the ^{10}Be exposure dating method. Furthermore, 5 former glacial stadials where reconstructed based on the mapped moraines and the Accumulation Area Ratio (AAR) method. For each stadial the Equilibrium Line Altitude and depressions in reference with the ELA from Little Ice Age are calculated.

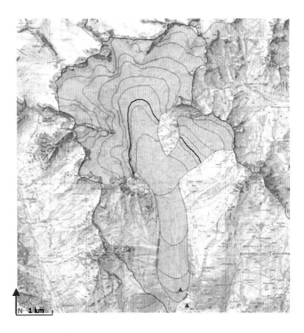

Fig. 2: *Glacier reconstruction based on dated boulders (red triangle) on Daun or Egesen stadial moraines (pink lines).*

In combination with the absolute ages of surface exposure of the sampled features, glacier reconstruction could be compared and, as far as possible, incorporated with the known climate history and glacial stratigraphy of the region. The reconstructed glacial stadials could be assigned to the Lateglacial stadials of Daun/Clavadel, Egesen and Kromer/Kartell. The result is a detailed story of the landscape evolution of Val Tuoi covering important steps in glacier and climate history of the Lateglacial and Holocene of the region.

[1] *Physical Geography, University of Zurich*

DEGLACIATION HISTORY OF THE SIMPLON PASS REGION

^{10}Be surface exposure dating of ice-molded bedrock surfaces

A. Dielforder[1], R. Hetzel[2], P.W. Kubik

The Simplon Pass (2005 m elevation) is a major mountain pass that connects the Rhône Valley (Switzerland) with the Alpine foreland in northern Italy. During the last glacial period the pass area was glaciated as documented by ice-molded bedrock surfaces, trimlines and moraines. To better understand the response of high Alpine glaciers to Lateglacial climate changes we constrained the deglaciation history of the Simplon Pass area by applying ^{10}Be surface exposure dating to well preserved ice-molded bedrock surfaces [1].

Fig. 1: *Northeastward view of the Simplon Pass area. The Simplon Pass depression is the relatively flat area between the pass and the small hill Gampisch. Sample sites are indicated by white dots. The location in Italy is not shown on this photograph.*

We collected samples from different geomorphic domains. Most samples were collected within the Simplon Pass depression (Fig.1) that is considered to have been glaciated during the Oldest Dryas stadial. In addition, we took two samples at higher elevation near the Staldhorn, whose peak and southeastern flank remained ice-free since the end of the Last Glacial Maximum (LGM). Furthermore we took two samples 25 km southeast of the Simplon Pass in northern Italy (500 m elevation) to date the retreat of glaciers in the southern foreland of the Alps.

The ^{10}Be ages from the different geomorphic domains are very consistent and define distinct age groups.

Samples from the Simplon Pass depression cluster tightly between $\approx$ 13.5 and $\approx$ 15.4 ka indicating that the area became ice-free at 14.1±0.4 ka (external error of mean age) [1]. This age constraint is interpreted to record the melting of high valley glaciers in the Simplon region during the warm Bølling-Allerød interstadial shortly after the Oldest Dryas.

The samples collected at higher elevations at the Staldhorn yield older ^{10}Be ages of 17.8±0.6 ka and 18.0±0.6 ka [1]. These ages likely reflect the initial downwasting of the LGM ice cap. Similar ^{10}Be ages of 17.7±0.9 ka and 16.1±0.6 ka were obtained from the two samples from northern Italy. In combination with well-dated paleoclimate records, our new age data suggest that during the deglaciation of the European Alps the decay of the LGM ice cap was approximately synchronous with the retreat of piedmont glaciers in the foreland and was followed by melting of high-altitude valley glaciers after the transition from the Oldest Dryas to the Bølling-Allerød, when mean annual temperatures rose rapidly by $\approx$ 3 °C [1].

[1] A. Dielforder and R. Hetzel, Quat. Sci. Rev. 84 (2014) 26

[1] *Geology, University of Bern*
[2] *Geology and Paleontology, University of Münster, Germany*

LGM ICE SURFACE DECAY SOUTH OF THE MT. BLANC

[10]Be surface exposure dating around Courmayeur (IT)

C. Wirsig, S. Ivy-Ochs, M. Christl, N. Akcar[1], J. Zasadni[2], P. Deline[3], C. Schlüchter[1]

Clear trimlines in the high granite walls of Val Ferret and Val Veny at the southern side of the Mt. Blanc massif bear witness of the more than 1 km of ice that filled the valleys during the Last Glacial Maximum (LGM).

Fig. 1: *Inset: location of study site in the Alps (©Wikipedia), A: sample sites marked on satellite image (©Swisstopo), LGM ice surface reconstruction of the same area as A [1].*

Fig. 2: *Glacially molded top of Aiguille du Châtelet. Four samples were collected on top of the narrow ridge crest.*

In order to determine the beginning of the lowering of this ice mass, we collected 12

bedrock samples for surface exposure dating at three different sites (Fig. 1).

At Aiguille du Châtelet (Figs. 2 and3) and Testa Bernada glacially molded bedrock surfaces are well-preserved. They were covered by only 100-200 m of ice during the LGM. Around La Saxe we identified and sampled a cluster of granitic Roche-Moutonée features.

Samples were collected at exposed locations where we consider deposition of sediment or ice highly unlikely. The [10]Be surface exposure ages to be determined in 2014 should therefore directly correspond to the beginning phase of ice decay after the LGM in the high Alps.

Fig. 3: *Aiguille du Châtelet, ridge marked in red.*

[1] A. Bini et al., Bundesamt für Landes-topografie Swisstopo (2009) ISBN 978-3-302-40049-5

[1] Geology, University of Bern
[2] Geology, Akademia Górniczo-Hutnicza, Krakow, Poland
[3] EDYTEM, Université de Savoie, Le Bourget du Lac, France

LATE WEICHSELIAN ICE SHEET HISTORY IN NW-SVALBARD

^{10}Be surface exposure dating of bedrock and boulders

E.F. Gjermundsen[1], J.P. Briner[2], N. Akçar[3], O. Salvigsen[4], P.W. Kubik, N. Gantert[1], A. Hormes[1]

The chronology and the configuration of the Svalbard Barents Sea Ice Sheet (SBSIS) during the Late Weichselian (LW) are based on few and geographically scattered data. Here, cosmogenic ^{10}Be dating of boulders (Fig. 1) and bedrock along with lithological investigations were used to reconstruct the configuration and deglaciation history of the LW ice sheet in NW Svalbard (NWS).

We show that the SBSIS started to thin earlier than previous reports have stated, but also document a late retreat from the distal (coastal) areas. Investigations of erratic boulders in the north and the south point to a local ice dome in NW Svalbard (Fig. 2). Our reconstruction fits well with the hypothesis of a complex multi-dome ice sheet configuration over Svalbard and the Barents Sea during the LW glaciation, with numerous drainage basins feeding fast ice streams, separated by slow flow, possibly cold-based, inter-ice stream areas [1, 2, 3, 4].

Fig. 1: *Sampling an erratic boulder in southern NW Svalbard.*

Erratic boulders deposited on intact pre-LW surfaces indicate that the ice in some places has been cold-based non-erosive. ^{10}Be ages of bedrock samples from vertical transects in mountainous terrain shows that the summits of

NWS could not have been covered by warm-based erosive ice during LW. However, we cannot exclude a cold-based non-erosive ice cover over the very highest summits in NWS during LW.

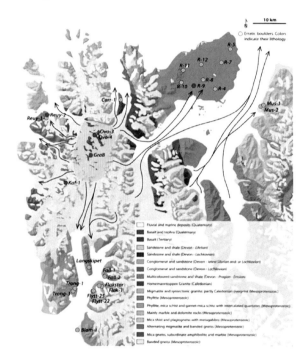

Fig. 2: *Geological map of NW Svalbard with the locations of rock samples along with the proposed ice dome and its drainage directions.*

[1]   J.Y. Landvik et al., Boreas 34 (2005) 7

[2]   D. Ottesen et al., Boreas 36 (2007) 286

[3]   D. Ottesen and J.A. Dowdeswell, GSA Bulletin 121 (11/12) (2009) 1647

[4]   H. Alexandersson et al., Boreas 40 (2011) 175

[1] *Arctic Geology, The University Centre in Svalbard, Longyearbyen, Norway*
[2] *Geology, University at Buffalo, USA*
[3] *Geology, University of Bern*
[4] *Geosciences, University of Oslo, Norway*

DEGLACIATION OF AN EAST ANTARCTIC OUTLET GLACIER

Exposure dating and modelling the retreat of the Skelton Glacier

R.S. Jones[1], K.P. Norton[1], A.N. Mackintosh[1], C.J. Fogwill[2], N.R. Golledge[1], P.W. Kubik

We investigate past glacier dynamics and evaluate surface exposure ages of Skelton Glacier, an outlet of the East Antarctic Ice Sheet. Initial ^{10}Be-drived exposure ages of erratic and bedrock samples, collected near the present-day ice surface in an upstream transect adjacent to the flow path, reveal a mixture of simple and complex exposures (Fig. 1).

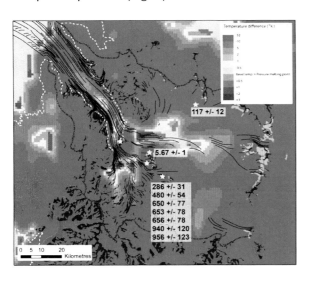

Fig. 1: *^{10}Be ages (ka BP) from samples on nunataks within Skelton Glacier's névé. Downstream samples indicate possible thicker ice at the LGM and mid-Holocene stabilization following thinning, and sufficient resetting.*

A flow-line model was used to recognize the dynamic and therefore potentially erosive areas of the glacier. The model was then run under a series of possible environmental conditions of the LGM (Fig. 2). We find that erratic samples showing complex exposure histories of repeated burial and cosmogenic inheritance can be explained by their distal proximity to both fast moving ice and ice at the pressure melting point, today and at the LGM, whereas erratics with simple exposure histories occur adjacent to and downstream from dynamic and potentially more erosive areas of the glacier.

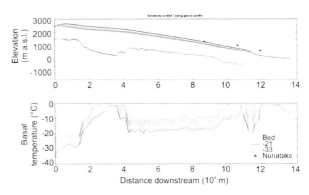

Fig. 2: *Exposure ages from nunataks combined with glacier flowline modelling to investigate both past dynamics of the glacier and the sample exposure histories.*

Using a combination of multiple isotope analysis and glacier flow-line modelling we conclude that (1) Skelton Glacier was thicker in the lower reaches but was unchanged or lower than present in the upper névé during the LGM, (2) lower reaches of the glacier stabilized at ca. 6 ka BP following deglaciation from the LGM extent, and (3) a more thorough consideration of the past and present glaciology of a catchment, including basal conditions and the transport path of an erratic, is necessary to obtain a reliable chronology of past ice sheet changes in Antarctica.

---

[1] *Geography, Victoria University of Wellington, New Zealand*

[2] *Biological, Earth and Environmental Sciences, University of New South Wales, Sydney, Australia*

LANDSCAPE EVOLUTION NORTH OF THE SONNBLICK

Surface exposure dating of a Egesen (YD) glacial system and landslides

M.G. Bichler[1], M. Reindl[1], J. M. Reitner[2], S. Ivy-Ochs, C. Wirsig, M. Christl

The area north of the Hoher Sonnblick peak in the Austrian province of Salzburg offers a great opportunity to study landscape forming events (glacial advances, glacial retreats and mass movements) since the Last Glacial Maximum.

Field work reveals temporally unique relationships of cross-cutting landscape elements. These include multiple moraines and a till cover of a dominant glacial stadial overlying a giant landslide ($\approx$ 0.4 km^3, largest in the province of Salzburg). On the other hand the basal till of this stadial itself is topped by a younger landslide of smaller dimension.

Fig. 1: *Dated features in the forefront of the (a) Little Ice Age terminal moraine of the Goldbergkees. (b) marks sampled boulders, bedrock and peatbogs on a high-plateau. (c) marks boulders on the left lateral moraine of the Egesen maximum extent and (d) are boulders representing the last retreat phase of the local Egesen spell.*

We obtained absolute geochronological data by surface exposure dating of 20 samples with the cosmogenic nuclide ^{10}Be (Figs. 1 and 2). The ^{10}Be ages were combined with a relative chronology based on field evidence as well as absolute ages by ^{14}C dating of the basal layers of peat, related to landslides and moraines. Thus, we obtained well constrained absolute ages of 2 landslides (13 ka (Fig. 2) and 10 ka BP) bracketing an Egesen (Younger Dryas) (12.5 - 10 ka BP) glacier system. In combination with a detailed geological and geomorphological map, it was possible to reconstruct the glacial chronology and the landscape evolution of the study area between 14 ka and 10 ka BP.

Fig. 2: *Dated boulder in the most distal part of the main landslide (13 ka) underlying Younger Dryas glacial deposits.*

We used various methods for calculating Equilibrium-Line-Altitudes and compared them to already available data from western Austria and Switzerland. With these data, we were able to reconstruct detailed aspects including temperature and precipitation change of the local climate and glacier dynamics during the maximum of the Younger Dryas for the first time in the European Alps east of the Brenner.

---

[1] *Geology, University of Vienna, Austria*
[2] *Geological Survey of Austria, Vienna, Austria*

THE AGE OF THE CASTELPIETRA ROCK AVALANCHE

Surface exposure dating of boulders with [36]Cl

S. Ivy-Ochs, S. Martin[1], A. Viganò[2], P. Campede[3], M. Rigo[1], V. Alfimov, C. Vockenhuber

As part of a project to assess paleoseismicity in the Trento Province of Italy, we are studying numerous rock avalanches in the Adige and Sarca River valleys. In the Adige Valley, between Trento and Rovereto, the Castelpietra and Lavini di Marco rock avalanche deposits are found on the left side of Adige. Based on the results of detailed mapping, we applied surface exposure dating to Lavini to determine the ages of the catastrophic events [1]. We also dated several mega boulders of the Castelpietra rock avalanche which lies about 10 km upstream from Lavini. The rock avalanche covers an area of about 0.7 km^2. With a short travel distance of only 0.7 km, a very steep Fahrböschung angle of 42° was calculated for the Castelpietra rock avalanche [2].

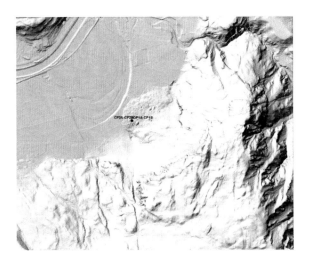

Fig. 1: *Scarp and deposits of the Castelpietra rock avalanche. Position of Dolomia Principale boulders exposure dated with [36]Cl.*

[36]Cl surface exposure dating of three Dolomia Principale mega boulder surfaces yielded consistent ages that seem to coincide with the timing of reactivation of the sliding planes at Lavini di Marco [1] about 1500 years ago. Ages were calculated using the production rates

listed in [3]. The early Medieval ages for reactivation at Lavini [1] may be contemporaneous with a catastrophic flood event (and related earthquake?) of the Adige river in Verona, as reported in the Fulda Annales, in 883 AD.

Fig. 2: *Scarp area and accumulation area of the Castelpietra rock avalanche. View to the east.*

The Castelpietra rock avalanche, as the Lavini reactivation, can be related to earthquakes. It has been proposed that some of the larger seismic events (e.g. 1117 AD in Verona and 1222 AD in Brescia) produced numerous rockslides in this region.

[1] S. Martin et al., Quat. Geochron. 19 (2014) 106

[2] G. Abele, Wissensch. Vereinshefte 25 (1974) 1

[3] V. Alfimov and S. Ivy-Ochs, Quat. Geochron. 4 (2009) 462

[1] *Geology, University of Padua, Italy*
[2] *Oceanography and Experimental Geophysics, CRS, Udine, Italy*
[3] *Geological Survey of the Province of Trento, Trento, Italy*

PREHISTORIC ROCK AVALANCHES IN THE OBERSEE AREA

Analysis, modeling and surface exposure dating (^{36}Cl) of two events

J. Nagelisen[1], J. Moore[2], C. Vockenhuber, S. Ivy-Ochs

A holistic analysis of two rock avalanches (Rautispitz and Platten) was made in the Obersee valley, Glarner Alps, Switzerland. It includes detailed mapping of all landslide and related Quaternary phenomena, new volume estimates for each event, runout modeling, and absolute dating of the rock slope failures.

Rautispitz (Fig. 1), the larger event, had a maximum horizontal travel distance of 5000 m, a fahrböschung angle of 18.4° and a volume of about 91 million m^3. The Platten rock avalanche (Fig. 2) had a maximum horizontal travel distance of 1600 m, a fahrböschung angle of 21.3° and a volume of about 11 million m^3. Both events have a calculated bulking factor around 28 %. They both failed along steeply-dipping bedding planes in Schrattenkalk limestone. Interspersed Orbitolina marl beds are assumed to have acted as basal rupture planes for the rock slope failures. Results of a runout analysis using the DAN3D landslide model showed an excellent match to mapped deposit extents and thickness distributions.

Fig. 1: *Release area and moving direction of the Rautispitz rock avalanche. View is to the south.*

The rock avalanches were dated using cosmogenic ^{36}Cl surface exposure dating. Thirteen boulders (8 for Rautispitz and 5 for Platten) spread over the deposit area were carefully selected and sampled. Results showed the Rautispitz rock avalanche to have occurred 11.6±0.8 ka BP. The Platten rock avalanche was found to have a mid-Holocene age of roughly 5.6±0.6 ka BP. Precise determination of individual landslide predisposing and triggering factors remains speculative.

Nonetheless, it is likely that the conditioning and preparatory factors for the Rautispitz rock avalanche were related to changing climate (warming after the Younger Dryas in combination with higher precipitation) and a change of rock slope geometry (undercutting during previous glacial advances), while the final trigger that released the rock avalanche may have been an earthquake.

The Platten rock avalanche was likely predisposed by erosion and weathering processes and finally the slope failed towards the end of the Holocene Climate Optimum after long-term progressive reduction of internal strength.

Fig. 2: *Release area, moving direction and deposit area of the Platten rock avalanche.* View is to the southwest with the town of Näfels in the foreground.

[1] *Geology, ETH Zurich*
[2] *Geology and Geophysics, University of Utah, Salt Lake City , USA*

MULTI-METHOD APPROACH TO DATE TSCHIRGANT

^{14}C, ^{36}Cl and ^{234}U/^{230}Th dating of a major prehistoric rockslide

M. Ostermann[1], S. Ivy-Ochs, C. Prager[2], D. Sanders[1]

The Tschirgant rockslide (Tyrol, Austria) is situated in the highly populated Upper Inn valley at the confluence between the Inn River, flowing towards the east, and the Ötz River, coming from the south. The terrain is typically very steep and rugged, with over-deepened valleys, filled with sediments from different glacial episodes and their following warm periods. The rockslide detached from a mountain flank more than 1400 m in vertical height, from an intensely deformed succession of Triassic dolostones and cellular dolomites, and left rock debris with a volume of about 230 million m^3 spread over an area of 9 km^2 (Fig. 2).

Fig. 2: *LIDAR image (TIRIS) of the Tschirgant rockslide area. Rock debris from the catastrophic rock slope failure is dyed in brown.*

In spite of a total of 17 numerical ages – a figure rarely produced for a catastrophic rock slope failure deposit – it proved not straightforward to deduce the time slot(s) of mass-wasting. Our results imply that event ages of catastrophic rock slope failures dated by a single age value, or a few values, may represent crude proxies only.

[1] M. Ostermann et al., Geomorph. 171-172 (2012) 83
[2] V. Alfimov and S. Ivy-Ochs, Quat. Geochron. (2009) 462

[1] *Geology, University of Innsbruck, Austria*
[2] *alpS Centre for Natural Hazard Management, Innsbruck, Austria*

Fig. 1: *Scarp area of the Tschirgant rockslide.*

The age of the event has been discussed since the first detailed investigations on the rockslide in the late 1960s. Nine already published radiocarbon dates have been recalculated and compared with radiometric ages derived from two other methods: (A) ^{234}U/^{230}Th dating of soda-straw stalactites formed in microcaves beneath rockslide boulders (three sites) and (B) ^{36}Cl cosmogenic surface exposure dating of boulders (five sites).

DATING SWISS DECKENSCHOTTER

Depth-profile and isochron-burial dating with cosmogenic nuclides

A. Claude[1], N. Akçar[1], S. Ivy-Ochs, P.W. Kubik, C. Vockenhuber, A. Dehnert[2], M. Rahn[2], C. Schlüchter[1]

Deckenschotter (cover gravels) are Quaternary sediments, which cover Tertiary Molasse or Mesozoic bedrock and are located beyond the limit of the Last Glacial Maximum. The Deckenschotter are a succession of proximal glaciofluvial gravels of the Northern Alpine Foreland, showing locally an intercalation with till and overbank deposits from (paleo-) valleys. These deposits, which can be differentiated by their distinct topographical position, are divided into two main geomorphic units: Höhere (Higher) and Tiefere (Lower) Deckenschotter. Even though the Höhere Deckenschotter occupies a topographically higher position, it is older than the Tiefere Deckenschotter as the two are separated from each other by a phase of incision. Both Höhere and Tiefere Deckenschotter bear evidence of at least four glacial advances that reached the Alpine foreland [1] and are, therefore, complex lithostratigraphic sequences. Reconstruction of the chronology of these glaciofluvial units will provide fundamental information about the onset of Quaternary glaciation in the Alps as well as about the timing and magnitude of incision on the foreland.

So far, we collected 54 samples from three sites (Fig. 1): Pratteln (project in collaboration with IPNA - Institut für Prähistorische und Naturwissenschaftliche Archäologie), Stadlerberg (Fig.2) and Irchel. The methods we apply are isochron-burial dating with ^{10}Be and ^{26}Al and depth-profile dating with ^{10}Be or ^{36}Cl including age modelling using Matlab codes [3, 4].

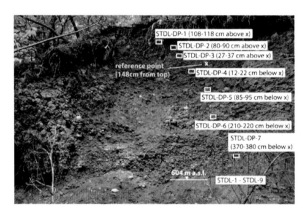

Fig. 2: *Sampling a 5 m long depth-profile at Stadlerberg. The red square shows the location of samples STDL-1 to STDL-9 for isochron dating.*

The depth-profile at Pratteln yielded a deposition age of around 300 ka and that at Stadlerberg of approximately 1.4 Ma.

[1] H.R. Graf, ETH Dissertation Nr. 10205 (1993)

[2] A. Bini et al., Bundesamt für Landestopografie Swisstopo (2009) ISBN 978-3-302-40049-5

[3] A.J. Hidy et al., Geochem. Geophys. Geosyst. 11 (2010) DOI: 10.1029/2010GC003084

[4] D. Tikhomirov et al., manuscript in prep.

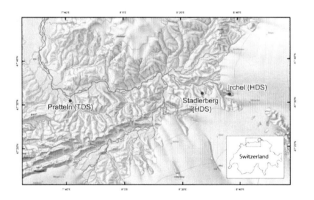

Fig. 1: *Location of the study sites on top of the map showing the Last Glacial Maximum (LGM) (after [2]).*

[1] *Geology, University of Bern*
[2] *Swiss Federal Nuclear Safety Inspectorate, ENSI, Brugg*

^{10}Be DEPTH PROFILES IN THE WESTERN SWISS LOWLANDS

Deposition ages of moraines are still difficult to constrain

L. Wüthrich[1, 2], R. Zech[1], N. Haghipour[1], C. Gnägi[3], M. Christl, S. Ivy-Ochs, H. Veit[2]

During the Pleistocene, glaciers advanced repeatedly from the Alps onto the Swiss foreland Molasse basin. The exact extent and timing are still under debate, even for the last glacial advances. Decalcification depths, for example, increase from west to east in the western Swiss lowlands and have been interpreted to indicate that the Valais (Rhone) glacier might have been most extensive asynchronous with the global Last Glacial Maximum (LGM) at 20 ka, but conversely occurred early during the last glacial cycle [1].

Fig. 1: *Sample locations for the Steinhof section.*

In an attempt to provide more quantitative age control, we applied ^{10}Be depth profile dating [2] on moraines at two locations. The first is a gravel pit near Niederbuchsiten, which presumably lies outside the LGM ice extent [1]. The second location, Steinhof, is located south of Herzogenbuchsee (Fig. 1). Here, two boulders were previously exposure dated and LGM ages were obtained [3]. Nevertheless, the decalcification depth at that site is unexpectedly high with 3.9 m. The ^{10}Be concentrations at both sites decrease with depth (Fig. 2).

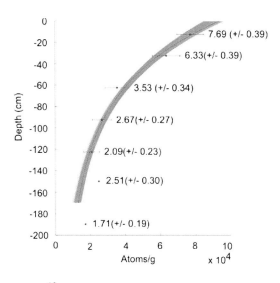

Fig. 2: *^{10}Be concentrations as a function of depth for the Steinhof section.*

The Matlab code of Hidy et al. [2] yields a deposition age of about 20 ka for Steinhof, consistent with the previously published boulder exposure ages [3]. The deposition age of the till at Niederbuchsiten cannot be constrained tightly. As exemplified at Niederbuchsiten, the lack of knowledge concerning the erosion history of the top surface poses a serious challenge for successfully application of depth-profile dating at foreland sites.

[1]　T. Bitterli et al., Geolog. Atlas d. Schweiz (2011) Blatt 1108

[2]　A.J. Hidy et al., Geochem. Geophys. Geosyst. 11 (2010) DOI: 10.1029/ 2010GC003084

[3]　S. Ivy-Ochs et al., Eclog. Geol. Helv. 97 (2004) 47

[1] *Geology, ETH Zurich*
[2] *Geography, University of Bern*
[3] *Herzogenbuchsee, Switzerland*

COSMOGENIC SEDIMENT FINGERPRINTING

The cosmogenic ^{10}Be and ^{14}C inventory of debris flow source areas

F. Kober [1], K. Hippe, B. Salcher [2], S. Ivy-Ochs, R. Grischott [3], L. Wacker, M. Christl, N. Hählen [4]

We investigated debris flow catchments in the Haslital Aare valley of Central Switzerland, from a sediment yield and a cosmogenic nuclide perspective. Localized mobilization of sediment as debris flows due to rockfall, heavy rainfall and permafrost thawing has been quantified volumetrically and in terms of cosmogenic nuclide (^{10}Be, ^{14}C) concentrations. Sediment sources and reservoirs (talus slope deposits, glacial debris, hillslopes, debris flow fans (Figs. 1 and 2)) are investigated at the source site, at the tributary - trunk stream junction (debris-flow subcatchment scale, ≈ 4 km^2) and at the outlet of the Haslital catchment (≈ 70 km^2).

Fig. 1: *The Spreitlaui debris flow catchment with sampled source areas. View to the west.*

The measurements indicate that the sediment is not as thoroughly mixed at the subcatchment and catchment scale as required by the concept of cosmogenic nuclide denudation rate estimations. A time series extended for two more years [1] confirms earlier results of a buffering of debris flow perturbation events in headwaters with larger catchment areas.

The combined analysis of the ratio of ^{10}Be and ^{14}C suggests that sediment storage can be neglected on the timescale resolved (a few hundred – to one thousand years).

Fig. 2: *The Rotlaui moraine bastion at about 2200 m with the sampled sediment sources.*

The analysis also reveals that the bulk of the material at the outlet of the subcatchments has cosmogenic nuclide concentrations equal to the upper and lower fan areas. For instance, in the Spreitlaui, the concentrations of high talus deposits are significantly higher than those of sediment from the fan area. Field and volumetric investigations confirm this obser-vation, where debris flows are triggered in high elevated parts of talus sheets or moraines, but the bulk of the material is entrained in the lower course.

Cosmogenic nuclide analysis in alpine catchments requires a thorough understanding of the processes occurring in the catchment and their temporal and spatial variation.

[1]   F. Kober et al., Geology 40 (2012) 935

[1] *Geology, ETH Zurich / Nagra, Wettingen*
[2] *Geology, University of Salzburg, Austria*
[3] *Geology, ETH Zurich*
[4] *Oberingenieurkreis Thun*

THE PROBLEM OF SEDIMENT MIXING

Pattern of sediment mixing revealed through concentrations of ^{10}Be

S. Savi[1], K.P. Norton[1], V. Picotti[2], F. Brardinoni[3], N. Akçar[1], P.W. Kubik, R. Delunel[1], F. Schlunegger[1]

Basin-wide erosion rates can be determined through the analysis of in situ-produced cosmogenic nuclides. In transient landscapes, and particularly in mountain catchments, erosion and transport processes are often highly variable and consequently the calculated erosion rates can be biased. This can be due to sediment pulses and poor mixing of alluvial sediment in the rivers. The mixing of alluvial sediment is one of the principle conditions that need to be verified to have reliable results.

We performed a field-based test [1] of the extent of sediment mixing for a ≈ 42 km^2 catchment in the Alps using concentrations of river-born ^{10}Be. We used this technique to assess the mechanisms and the spatio-temporal scales for the mixing of sediment derived from hillslopes and tributary channels. The results show that sediment origin and transport, and mixing processes have a substantial impact on the ^{10}Be concentrations downstream of the confluence between streams. We also illustrate that the extent of mixing significantly depends on: the sizes of the catchments involved, the magnitude of the sediment delivery processes, the downstream distance of a sample site after a confluence, and the time since the event occurred. In particular, soil creep and shallow-seated landsliding supply high ^{10}Be concentration material from the hillslope, congruently increasing the ^{10}Be concentrations in the alluvial sediment (till and soil contribution in Fig. 1). Contrariwise, a high frequency of mass-wasting processes or the occurrence of sporadic but large-magnitude events (debris flow contribution in Fig. 1) results in the supply of low-concentration sediment that lowers the cosmogenic nuclide concentration in the channels. The predominance of mass-wasting processes in a catchment can cause a strong bias in detrital cosmogenic nuclide concentrations, and therefore calculated

erosion rates may be significantly over- or underestimated. Accordingly, it is important to sample as close as possible to the return-period of large-size events. This will lead to an erosion rate representative of the "mass-wasting signal" in case of generally high-frequency events, or the "background signal" when the event is episodic.

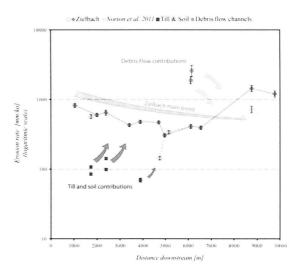

Fig. 1: *Pattern of downstream sediment mixing at Zielbach, South Tyrol, taken from [1].*

Our results suggest that a careful consideration of the extent of mixing of alluvial sediment is of primary importance for the correct estimation of ^{10}Be-based erosion rates in mountain catchments, and likewise, that erosion rates have to be interpreted cautiously when the mixing conditions are unknown or mixing has not been achieved.

[1] S. Savi et al., Quat. Geochron. 19 (2014) 148

[1] *Geology, University of Bern*
[2] *Geology, University of Bologna, Italy*
[3] *Geology, University of Milano-Bicocca, Italy*

THE EROSION RATE PATTERN AT NIESEN

Erosion rate determination through [10]Be

H. Chittenden[1], R.Delunel[1], F. Schlunegger[1], N. Akçar[1], P.W. Kubik

Landscape evolution and surface morphology in mountainous settings are a function of the relative importance between sediment transport processes acting on hillslopes and in channels, modulated by climate variables. The Niesen nappe in the Swiss Penninic Prealps (Fig. 1) presents a unique setting, in which opposite facing flanks host basins underlain by identical lithologies but contrasting litho-tectonic architectures where lithologies either dip parallel to the topographic slope or in the opposite direction (i.e. dip slope and non-dip slope).

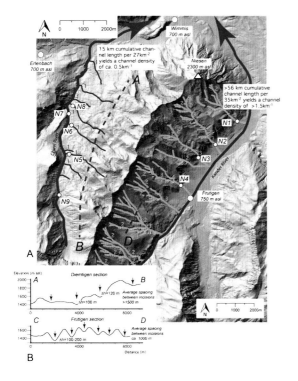

Fig. 1: *Morphology of the Niesen, and sites (N1-N9) where modern stream sediments for [10]Be analyses have been collected.*

The northwest facing Diemtigen flank represents such a dip slope situation and is characterized by a gentle topography, low hillslope gradients, poorly dissected channels, and it hosts large landslides. In contrast, the southeast facing Frutigen side can be described as a non-dip slope flank with deeply incised bedrock channels, high mean hillslope gradients and high relief topography [1].

While the contrasting dip-orientations of the underlying flysch bedrock have promoted hillslope and channelized processes to contrasting extents and particularly the occurrence of large landslides on the dip slope flank, the flank averaged [10]Be-derived denudation rates are very similar and range between 0.20 and 0.26 mm/a. In addition, our denudation rates offer no direct relationship to basin slope, area, steepness or concavity index, but reveal a positive correlation to mean basin elevation that we interpret as having been controlled by climatically driven factors such as frost-induced processes and orographic precipitation [1].

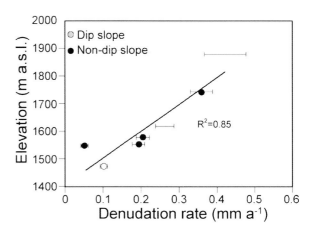

Fig. 2: *Relationships between [10]Be-based denudation rates and mean elevations of analyzed catchments.*

[1] H. Chittenden et al., Earth Surf. Process. Landf. (2013) DOI: 10.1002/esp.3511

[1] *Geology, University of Bern*

TEMPORAL STEADY-STATE INCISION (ONSHORE MAKRAN)

Surface exposure dating of fluvial terraces with ^{10}Be

N. Haghipour[1], J-P. Burg[1], S. Ivy-Ochs, P.W. Kubik, M. Christl

Most of the studies on Quaternary deposits in eastern and central Iran are limited to Holocene [1]. Several sets of fluvial terraces are preserved along the main valleys in Makran (Fig. 1).

Fig. 1: *Photographs of fluvial terraces (T1 – T4) inland Makran.*

Thirty-five strath terraces have been dated using *in situ* produced cosmogenic ^{10}Be concentrations from surfaces and depth profiles [1]. These new measurements yield abandonment ages between ≈ 13 and 379 ka. The age distribution allows determining the chronology of terrace levels and establishing regional correlations between deposits of four large

catchments (Fig. 2). The comparison of incision rates in these catchments enables distinguishing between a relatively moderate regional, "background" incision rate of ca 0.3±0.04 mm/a and local, tectonic-driven incision/uplift (0.8-1±0.1 mm/a) rates. The similarity between regional fluvial incision rates (0.3-0.4 mm/a) and Pleistocene coastal uplift rates constrained by the newly dated marine terraces (0.2 mm/a) supports the initial working hypothesis that fluvial rivers responded to a regional, long-term interplay between climatically-driven incision and tectonically-driven surface uplift. The obtained uniform incision rate of 0.3 mm/a indicates tectonic steady-state of the wedge on a regional scale. However, perfect steady-state is unlikely on short-length scales.

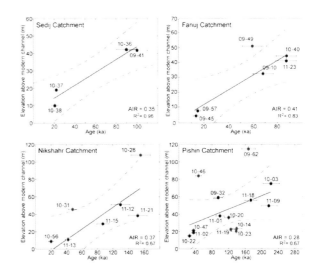

Fig. 2: *Incision plots for studied catchments. AIR = apparent incision rate*

[1]   R.T. Walker et al., Quat. Sci. Rev. 30 (2011) 1256

[2]   N. Haghipour et al., Earth Planet. Sci. Lett. 355-356 (2012) 187

[1] *Geology, ETH Zurich*

EROSION DISTRIBUTION IN THE EASTERN HIMALAYA

[10]Be denudation in the Brahmaputra basin (Tibet, India & Bangladesh)

M. Lupker[1], J. Lavé[2], C. France-Lanord[2], P.W. Kubik, D. Bourlès[3], R. Wieler[1]

The Brahmaputra River flows from the high and arid Tibetan Plateau through the dissected Himalayan range and further over a flat and humid floodplain to the Bay of Bengal (Fig. 1). It exports ca. 600 Mt of sediments per year to the ocean.

Intense fluvial incision and erosion along the Brahmaputra in the eastern Himalaya syntaxis has been suggested to drive high uplift rates in the region through lithospheric unloading [1]. Despite being a potential key region to better understand the couplings between erosion, tectonics and climate, modern denudation in the Brahmaputra watershed is still poorly quantified.

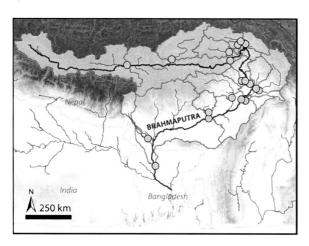

Fig. 1: *River sediment sample locations.*

To address these questions we measured cosmogenic [10]Be on river sediments along the entire length of the river (Fig. 1) to retrieve modern erosion rates and sediment fluxes. These measurements suggest low denudation rates on the Tibetan Plateau (<0.1 mm/a) that increase locally to 2 to 3 mm/a in the region of the eastern syntaxes of the Himalayan range. These samples also highlight the variability of the cosmogenic signal, since different sampling seasons or grain sizes show variable [10]Be

concentrations by up to a factor 3 for a same location. This variability is attributed to the poor mixing of sediments in this highly stochastic environment and makes the reconstruction of sediment fluxes more difficult.

On-going work aims to link the variations of [10]Be concentration to changes in sediment source. Recent measurements of the oxygen isotopic composition of individual quartz grains ($\delta^{18}O$) as a provenance proxy on the same samples may contribute to a more robust interpretation of [10]Be data in terms of erosion rates and sediment fluxes (Fig. 2).

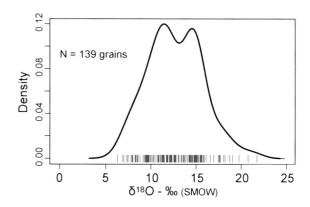

Fig. 2: *Oxygen isotopic composition of quartz from Brahmaputra sand sampled in Bangladesh.*

[1] N. Finnegan et al., Geol. Soc. Am. Bull. 120 (2008) 142

[1] *Geochemistry and Petrology, ETH Zurich*
[2] *Centre de Recherches Pétrographiques et Géochimiques (CRPG), Nancy, France*
[3] *Centre Européen de Recherche et d'Enseignement des Géosciences de l'Environement (CEREGE), Aix-en-Provence, France*

CONTRASTING EROSION IN DISTURBED ECOSYSTEMS

Impact of humans on erosion processes is ecosystem-dependent

V. Vanacker[1], N. Bellin[1], J. Schoonejans[1], A. Molina[2], P.W. Kubik

Very few, if any, mountain ecosystems remain unaffected by human disturbances. Here, we present empirical data that allow evaluating changes in erosion rates after human disturbances. Pre-disturbance (or natural) erosion rates are derived from in-situ produced [10]Be concentrations in river sediment, while post-disturbance (or modern) erosion rates are derived from sedimentation rates in small catchments. Data are presented for two mountain ecosystems located in the semi-arid Betic Cordillera (Spain) and the Tropical Andes (Ecuador).

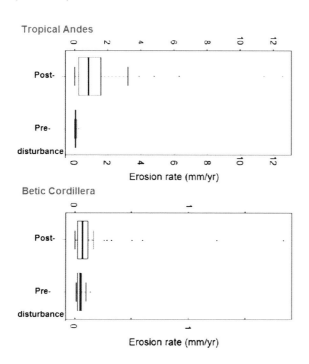

Fig. 1: *Boxplots of pre- and post-disturbance erosion rates for the Tropical Andes and the Betic Cordillera.*

Our data show evidence that modern erosion rates are not necessarily equivalent to human-induced erosion rates, as natural erosion rates can be important in mountainous terrain.

While the Spanish Betic Cordillera is commonly characterized as a degraded landscape, there is no significant difference between modern catchment-wide erosion and long-term denudation rates. The opposite is true for the Tropical Andes where the share of natural erosion in the total modern erosion rate is minimal for most disturbed sites.

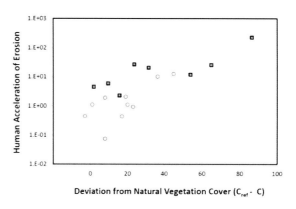

Fig. 2: *Human acceleration of erosion as function of vegetation disturbance for Tropical Andes (rectangles) and Spanish Cordillera (dots).*

When pooling pre- and post-disturbance erosion data, we observe that the human acceleration of erosion (defined as ratio of post- to pre-disturbance erosion rate), is an exponential function of vegetation disturbance. This suggests that the sensitivity to human-accelerated erosion might be ecosystem dependent, where well-vegetated ecosystems are more sensitive due to their greater exposure to strong vegetation disturbances.

[1] V. Vanacker et al., Land. Ecol. (2014) in press

[2] N. Bellin et al., Quat. Int. 308 (2013) 112

[1] *Earth & Life Institute, University of Louvain, Belgium*
[2] *PROMAS, Universidad de Cuenca, Cuenca, Ecuador*

ANTHROPOGENIC RADIONUCLIDES

AMS of americium and curium

Ultra-sensitive detection of actinides

Automated urine analysis method

Sequential extraction of ^{137}Cs, ^{90}Sr and actinides

^{236}U and ^{129}I in the Arctic Ocean

Mediterranean GEOTRACES

^{236}U and Pu in corals from French Polynesia

Determination of plutonium in Japanese samples

Iodine-129 in aerosols from Germany

^{129}I in exhaust air of German nuclear facilities

Long-lived halogen nuclides in absorber foils

AMS OF AMERICIUM AND CURIUM

A sensitivity study for the higher actinides at low energies

M. Christl, X. Dai[1], J. Lachner, S. Kramer-Tremblay[1], H.-A. Synal

Accelerator mass spectrometry (AMS) is one of the most sensitive, selective, and robust techniques for actinide analyses. While measurements of U and Pu isotopes have become routine at the ETH Zurich AMS system TANDY, there is an increasing demand for highly sensitive analyses of the higher actinides such as Am and Cm for bioassay applications. In order to extend the actinide capabilities of the TANDY system and to develop new, more sensitive bioassay routines, a pilot study was carried out in collaboration with Chalk River Labs (Canada). The aim of this study was to investigate and document the performance and the potential background of Am and Cm isotopic analyses with low energy AMS.

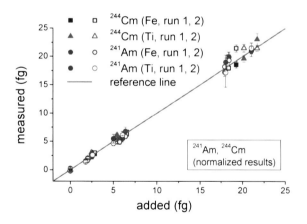

Fig. 1: *Measured vs added amount of ^{241}Am and ^{244}Cm after normalization. Brown and red (blue and lilac) indicate samples prepared in Fe (Ti) matrix. Same symbol types indicate identical samples measured in different AMS runs.*

Since a sufficiently pure Cm spike material was not available, all isotopes were measured relative to a ^{243}Am spike. To investigate sample matrix effects and to determine differences in the negative ion yield of Am vs Cm, the samples were prepared in Fe plus Nb and in Ti plus Nb matrices, respectively.

Our results show that Cm isotopes can only be determined relative to ^{243}Am, if samples and AMS standards are prepared identically with regard to the matrix elements, in which the sample is dispersed [1]. When corrected for these matrix effects the measured values are in good agreement with the added amounts of Am and Cm (Fig. 1). In our first test, detection limits for Cm isotopes are all below 100 ag (Tab. 1). The slightly higher value for ^{241}Am can be explained by ^{241}Am contained in the ^{243}Am spike.

Detection limits	[ag]	[at]	[Bq]
^{244}Cm	17	4×10^4	5×10^{-5}
^{246}Cm	41	1×10^5	5×10^{-7}
^{247}Cm	48	1×10^5	2×10^{-10}
^{248}Cm	63	2×10^5	1×10^{-8}
^{250}Cm	48	1×10^5	3×10^{-7}
^{241}Am	121	3×10^5	2×10^{-5}

Tab. 1: *Detection limits for Am and Cm isotopes given in attogram (ag), number of atoms and activity [1].*

In a systematic background study, two formerly unknown metastable triply charged Th molecules were found at masses 244 (ThC^{3+}) and 248 (ThO^{3+}). The presence of such a background is not a principal problem for AMS, if the stripper pressure is adjusted high enough to ensure destruction of all molecular interferences. We conclude that ultra-trace analyses of Am and Cm isotopes for bioassay are very well possible with low energy AMS.

[1]   M. Christl et al., Nucl. Instr. & Meth. (2014) accepted

[1] *Chalk River Nuclear Laboratory, Atomic Energy of Canada Limited, Chalk River, Canada*

ULTRA-SENSITIVE DETECTION OF ACTINIDES

Improved target preparation method for Th, U, Np, Pu, Am, Cm and Cf

X. Dai[1], M. Christl, S. Kramer-Tremblay[1], J. Lachner, H.-A. Synal

There is increasing demand for highly sensitive measurements of the higher actinides (Am, Cm and Cf) in environmental and biological samples as these transplutonic elements can accumulate gradually in a reactor through continuous neutron capture reactions. Due to their high radiotoxicity and intermediate half-lives, Am and Cm isotopes (e.g., ^{241}Am and ^{244}Cm) are often found to be the most significant dose contributors for possible internal exposure at reactor power plants. The most commonly used urine bioassays by alpha spectrometry often do not meet the sensitivity requirements for these nuclides. On the other hand, as a means of nuclear forensics, accurate and ultra-sensitive determination of isotopic signature of the higher actinides is very important to track down the source origin of radioactive materials. These applications would require the most sensitive technique, such as AMS, to be utilized for the detection of actinide isotopes at femtograms (fg) or even lower levels.

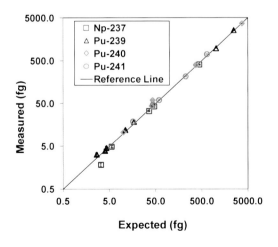

Fig. 1: *Measured vs. expected Pu and Np isotopes (using a ^{242}Pu tracer) mixed with titanium and iron oxide.*

To evaluate the performance of Compact AMS, samples were spiked with fg-levels of actinides (including Th, U, Np, Pu, Am, Cm and Cf isotopes) and measured at the TANDY AMS system. As exemplarily illustrated in Fig. 1 and 2, good agreement between the measured and expected values at the fg range has been achieved for Pu, Np, and Cf (but also for Th, U, Am and Cf isotopes, not shown here).

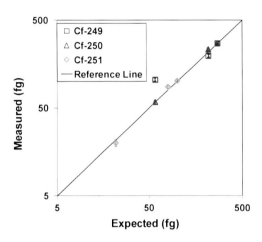

Fig. 2: *Measured vs. expected Cf isotopes (using a ^{252}Cf tracer) mixed with titanium and iron oxide.*

High and consistent ionization efficiencies were observed for all the actinides. Due to very low background, detection of attogram levels of Cm and Am isotopes could be demonstrated [2]. The compact ETH AMS system outperformed much larger traditional AMS facilities in ultra-trace actinide analysis and thus becomes a very competitive method for routine radioassays.

[1] X. Dai et al., J. Anal. Atom. Spectrom. 27 (2012) 126

[2] M. Christl et al., Nucl. Instr. & Meth. B (2014) accepted

[1] *Chalk River Nuclear Laboratory, Atomic Energy of Canada Limited, Chalk River, Canada*

AUTOMATED URINE ANALYSIS METHOD

Simultaneous determination of ultra-trace plutonium and neptunium

J. Qiao [1], X. Hou [1,2], P. Roos [1], J. Lachner, M. Christl, Y. Xu [1,3]

Neptunium (Np) and plutonium (Pu) are regarded as highly radiologically and biologically toxic radionuclides. Assessment of the exposure level to Np and Pu is required for radiation protection and medical intervention whenever people are exposed to a Np and Pu contaminated environment [1].

We developed a sequential injection-extraction chromatographic (SI-EC) method for simultaneous determination of Np and Pu with accelerator mass spectrometry (AMS) in 0.2 - 1 L of urine (Fig. 1).

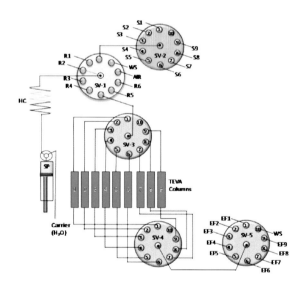

Fig. 1: *Schematic demonstration of the SI-EC system for Np and Pu determination (SP: syringe pump; HC: holding coil; S1-S9: ports for sample loading; FE1-EF9: ports for eluate collection; WS: waste; SV1-SV5: selective valves; R1-R6: reagents for column separation)*

Iron hydroxide co-precipitation was used in the pre-concentration step. Crucial experimental parameters affecting the analytical performance were investigated in detail.

For this method, ^{242}Pu performed well as a chemical yield tracer for both Pu and Np

isotopes. It also serves as a spike for the AMS measurement of ^{239}Pu and ^{237}Np.

The measurement results of urine samples spiked with ^{239}Pu and ^{237}Np agree well with the expected values for ^{239}Pu down into the single fg level and show a reasonable agreement for ^{237}Np in the range above 10 fg.

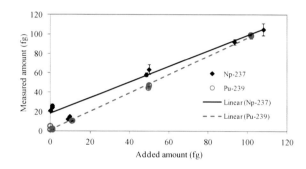

Fig. 2: *Measured versus added ^{237}Np and ^{239}Pu amounts in 200 mL urine samples.*

Linear fits (y=a·x+b) result in parameters of a=0.80±0.06, b=18±3 for ^{237}Np and a=0.95±0.02, b=1.0±0.9 for ^{239}Pu, respectively (Fig. 2). The quality of the linear fits, with correlation coefficients (R^2) of 0.955 and 0.997 for ^{237}Np and ^{239}Pu, respectively, documents the reliability of the proposed method. The limit of detection (LOD) for ^{239}Pu was calculated to be 3 fg. The relatively high ^{237}Np blank level (18±3 fg) increasing the LOD of ^{237}Np to ≈ 20 fg needs to be investigated further.

[1]   J. Qiao et al., Anal. Chim. Acta 652 (2009) 66

[1] *Nutech, Technical University of Denmark, Roskilde, Denmark*

[2] *Earth Environment, Chinese Academy of Sciences, Xi'an, China*

[3] *Coastal and Island Development, Nanjing University, China*

SEQUENTIAL EXTRACTION OF ^{137}Cs, ^{90}Sr AND ACTINIDES

How to get ^{137}Cs, ^{90}Sr and actinides from one single seawater sample?

N. Casacuberta, M. Castrillejo[1], P. Masqué[1], M. Christl, C. Breier[2], S. Pike[2], K. O. Buesseler[2]

One of the big issues when sampling seawater for the analysis of anthropogenic radionuclides such as ^{137}Cs and ^{90}Sr is the large amount of water (20 - 100 L) needed for quantification. Here, we present a method developed in collaboration with Universitat Autònoma de Barcelona and Woods Hole Oceanographic Institution (WHOI) for the extraction of ^{137}Cs, ^{236}U and ^{90}Sr from a single 20 L seawater sample.

The samples were filtered on board using a Hytrex cartridge (pore size 1 μm) and acidified to pH = 1. Stable ^{133}Cs (0.7 mg), ^{233}U (5 pg) and natural Sr (200 mg) were added to the samples in order to quantify chemical recovery.

For the Cs separation a composite inorganic ion exchanger was prepared at the Czech Technical University in Prague [1] by incorporating potassium-nickel hexacyanoferrate (II) (KNiFC) into a binding matrix of modified poly-acrylonitrile (PAN) [1]. Samples were passed through 5 mL of ion exchanger (Fig. 1), and then dried and transferred to polyethylene containers for γ-counting (coaxial HPGe detector) at WHOI [2].

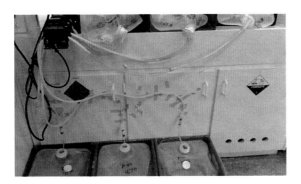

Fig. 1: *Samples flowing through KNiFC-PAN resin for the separation of Cs isotopes.*

Once the water has gone through the KNiFC-PAN column, a Fe^{2+} solution is added (200 mg). Iron hydroxides are formed after raising the pH to 8.5 with ammonia (Fig. 2). Actinides are scavenged from water and pre-concentrated in 250 mL bottles for further radiochemical purification and AMS analysis of U- and Pu-isotopes at ETH Zurich [3].

Fig. 2: *Pre-concentration of the samples for further processing and analysis of ^{236}U and Pu (iron hydroxides) and ^{90}Sr (supernatant).*

The remaining 20 L supernatant, free of Cs and actinides, is kept for the further analysis of ^{90}Sr. As ^{90}Sr is analyzed through its daughter (^{90}Y), samples should be stored for at least 18 days in order to reach secular equilibrium between ^{90}Sr and ^{90}Y (Y isotopes were also scavenged with the iron precipitates in the step before). The method used for its purification and quantification is based on [4].

[1] F. Sebesta, J. Radioanal. Nucl. Chem. 220 (1997) 77

[2] J. Kameník et al., Biogeosci. 10 (2013) 6045

[3] M. Christl et al., Nucl. Instr. & Meth. B 294 (2013) 29

[4] J.T. Waples and K.A. Orlandini, Limnol. & Oceanogr.: Methods 8 (2010) 661

[1] *Environmental Sciences, Universitat Autònoma de Barcelona, Spain*
[2] *Woods Hole Oceanographic Institution, Woods Hole, USA*

^{236}U AND ^{129}I IN THE ARCTIC OCEAN

Tracing the Atlantic waters into the Arctic Ocean

N. Casacuberta, M. Christl, C. Vockenhuber, C. Walther[1], M.R. Van-der-Loeff[2], P. Masqué[3], H.-A. Synal

Anthropogenic ^{236}U ($T_{1/2}$ = 23x10^6 a) has attested to be a new transient oceanographic tracer: it is conservative in seawater and far from having reached steady state in the oceans [1]. The main sources of ^{236}U in the North Atlantic Ocean are global nuclear fallout (ca. 1000 kg) and the European reprocessing plants of Sellafield and La Hague (ca. 120 kg) [2, 3].

In this study, ^{236}U and ^{129}I concentrations were determined in 8 water depth profiles in the Arctic Ocean collected in 2011/2012 (Fig. 1). Samples were measured with a compact ETH Zurich AMS system TANDY.

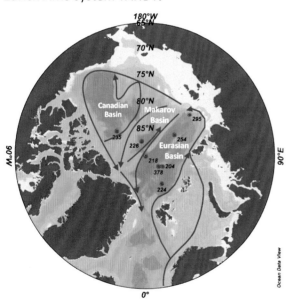

Fig. 1: *Location of the samples collected (red dots). Blue arrows represent the Atlantic water circulation.*

The results of the ^{236}U/^{238}U measurements show a steep gradient, from the lowest ever-reported ratio in open ocean (5±5)x10^{-12} up to (3700±80)x10^{-12}. Low values come from to the deep waters in the Canadian Basin and Makarov Basin stations (226 and 235), while the highest ones originate from the surface waters in the Eurasian Basin (Fig. 2). Similar trends are observed for ^{129}I concentrations, ranging from (4710±87)x10^6 to (1.6±0.8)x10^6 atoms/liter.

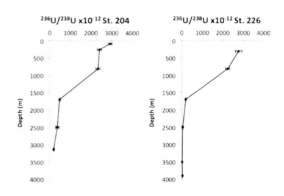

Fig. 2: *Profiles of ^{236}U/^{238}U ratios from station 204 (Eurasian Basin) and station 226 (Makarov Basin).*

Whereas the occurrence of ^{236}U can be attributed to both global fallout and nuclear reprocessing plants, the ^{129}I is mostly coming from the latter source. Therefore, the presence of both ^{236}U and ^{129}I traces the penetration of Atlantic waters into the Arctic Ocean following the cyclonic Atlantic Ocean Boundary Current (blue arrows in Fig. 1). The atomic ratio of ^{129}I/^{236}U might become a powerful tool to trace circulation time scales for Atlantic water in the Arctic Ocean.

[1] N. Casacuberta et al., Geochim. Cosmochim. Acta (2014) submitted

[2] P. Steier et al., Nucl. Instr & Meth. B 266 (2008) 2246

[3] A. Sakaguchi et al., Earth and Planet. Sci. Lett. 333-334 (2012) 165

[1] *Radioecology and Radiation Protection, University of Hannover, Germany*
[2] *Alfred Wegener Institute, Bremerhaven, Germany*
[3] *Environmental Sciences, Universitat Autònoma de Barcelona, Spain*

MEDITERRANEAN GEOTRACES

Tracing water masses with anthropogenic radionuclides

J. Garcia-Orellana[1], P. Masqué[1], M. Castrillejo[1], M. Roca-Martí[1], N. Casacuberta, M. Christl

In May 2013, the research vessel (R/V) *Ángeles de Alvariño* (Fig. 1) cruised the Mediterranean Sea as part of the European project *Mediterranean Sea Acidification in a changing climate (MedSeA)* and the GEOTRACES program. GEOTRACES aims to improve the understanding of bio-geochemical cycles and large-scale distribution of trace elements and their isotopes in the marine environment. The cruise was part of the GEOTRACES program to cover the section GA04, in combination of the Royal Institute for Sea Research (NIOZ) cruise on board the R/V *Pelagia*. Both cruises aimed to determine the distribution of the trace elements and isotopes selected by GEOTRACES and many other trace elements. *Pelagia* covered the trace metal-clean sampling, and *Ángeles Alvariño* collected those samples that required large volumes of water (i.e. samples for further analysis of anthropogenic radionuclides).

Fig. 1: *R/V Ángeles de Alvariño.*

Several anthropogenic radionuclides such as ^{14}C, ^{3}H, ^{137}Cs, ^{90}Sr, ^{236}U, ^{129}I, etc. are used in oceanography as powerful tools to quantify oceanic processes like water mass mixing, deep water formation rates, or to determine the age of water masses.

The distribution of anthropogenic radionuclides in the Mediterranean Sea is not well known. One aim of this cruise was therefore:

i) to study the concentration of ^{137}Cs, ^{90}Sr, Pu isotopes, ^{237}Np, ^{236}U and ^{129}I in different profiles collected across the Mediterranean Sea (Fig. 2),

ii) to constrain the sources of these radionuclides in this ocean (i.e. global fallout, Chernobyl accident and nuclear reprocessing plants),

iii) to trace oceanographic processes by combining the different behaviors of several anthropogenic radionuclides.

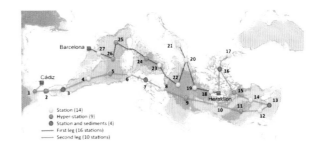

Fig. 2: *Track of the MedSeA cruise.*

A total of 70 samples from 10 deep profiles were collected for anthropogenic radionuclides analysis during the 5-week cruise. ^{137}Cs, ^{90}Sr, Pu isotopes and ^{237}Np will be processed at UAB (Barcelona), while ^{236}U and ^{129}I will be processed and analyzed at ETH Zurich.

[1] *Environmental Sciences, Universitat Autònoma de Barcelona, Spain*

^{236}U AND Pu IN CORALS FROM FRENCH POLYNESIA

First coral data from the Southern Hemisphere

M. Christl, B. Chayeron[1], N. Casacuberta

From 1966 to 1974, France conducted atmospheric nuclear bomb tests at the atolls of Mururoa and Fangataufa in French Polynesia, South Pacific Ocean (Fig. 1). During that period, also safety experiments were performed on the coral bedrock in the North of Mururoa which generated local Pu contamination in the lagoon.

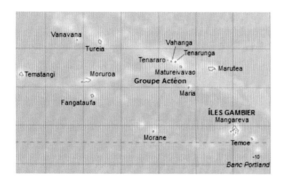

Fig. 1: Map of French Polynesia showing the French test sites and the sampling location (Gambier Islands).

To investigate the regional impact of these tests two coral samples A and B (Fig. 2 shows one of them) were collected on Gambier Islands about 500 km distant from the test sites. The islands have been exposed to radioactive fallout during the testing.

Fig. 2: Porites coral collected in 2012 in Gambier Islands.

Due to biological activity of the so-called coral borer lithophaga nigra the quality of both coral

samples is poor. The temporal resolution of the samples thus is only 3 to 7 years per sample with an unknown but probably large absolute age uncertainty.

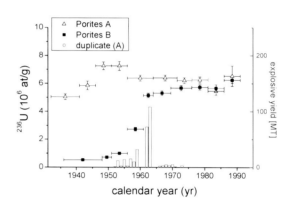

Fig.3: The ^{236}U concentrations in Porites A and B (left y-axis) and the estimated global yield of annual atmospheric nuclear bomb tests (right y axis) are plotted against the modelled age of the corals.

The ^{236}U concentrations in Porites B qualitatively agree with the history of atmospheric bomb testing (Fig. 3). The ^{236}U signal in Porites A, however, is clearly discordant for the older part of the coral. The elevated ^{236}U levels in Porites A (pre 1960) could have been caused by biological induced mixing or point to a severe problem with the chronology of this specimen.

Our results for ^{239}Pu/^{240}Pu show no clear trend but generally agree with data from the literature [1]. Measured ratios range between 0.2 and 0.05 indicating global fallout and weapons grade plutonium, respectively.

[1] R. Chiappini et al., Sci. Total Environ. 269 (1999) 237

---

[1] CRIIRAD Laboratory, Valence, France

DETERMINATION OF PLUTONIUM IN JAPANESE SAMPLES

Plutonium release from the Fukushima Daiichi nuclear power plant

S. Schneider[1], C. Walther[1], S. Bister[1], V. Schauer[2], M. Christl, H.-A. Synal, K. Shozugawa[3], G. Steinhauser[4]

During the Fukushima Daiichi nuclear power plant (FDNPP) accident, a large amount of radionuclides was released to the environment, but only small releases of actinides such as plutonium were expected. Since the time of the atmospheric nuclear weapons tests, plutonium is still omnipresent in the environment. Isotopic ratios can be used to differentiate between global fallout plutonium (^{240}Pu/^{239}Pu ≈ 0.18) and reactor plutonium (^{240}Pu/^{239}Pu ≈ 0.4 - 0.6). For FDNPP the isotopic ratio ^{240}Pu/^{239}Pu is approximately 0.4.

In 2011, soil and plant samples were taken from several Japanese hot spots (Fig. 1) and investigated using AMS and alpha-spectrometry [1]. Because the samples were rather small, alpha-spectrometry produced only two results (Tab. 1). Also, only three out of the 20 samples investigated yielded AMS results of detectable plutonium (according to ISO 11929) (Tab.1).

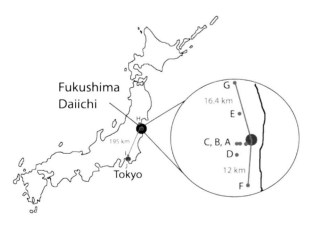

Fig. 1: *Locations of the investigated samples.*

The ^{240}Pu/^{239}Pu ratios of the three successfully measured samples are listed in Table 1. One sample (soil sample H-S2) shows an isotopic ratio typical for global fallout. The higher isotopic ratio of the sample A-V (plant sample) proves the release from a nuclear reactor. Though the ratio of 0.64 for sample G-V also

points at reactor plutonium, the high analytical uncertainty calls for careful interpretation. Nevertheless, the release of plutonium from the damaged FDNPP was demonstrated for at least one sample.

Sample name	^{240}Pu/^{239}Pu at/at	$^{239+240}$Pu Bq/kg
A-V	0.381 ± 0.046	0.49 ± 0.03
G-V	0.64 ± 0.37	0.17 ± 0.05
H-S2	0.205 ± 0.039	---

Tab. 1: *Measured ^{240}Pu/^{239}Pu ratios and Pu activity concentrations [1].*

The measured $^{239+240}$Pu activity concentrations are relatively low, which indicates a low plutonium release from Fukushima. For sampling locations close to each other, like soil and the plant growing on it, the plutonium content and the isotope ratios differ considerably. This strong localization indicates a particulate Pu release, which is of high radiological risk, if incorporated.

[1]   S. Schneider et al., Sci. Rep. 3, 2988 (2013) DOI:10.1038/srep02988

[1] *Radioecology and Radiation Protection , University of Hannover, Germany*
[2] *Atominstitut, Vienna University of Technology, Austria*
[3] *Graduate School of Arts and Sciences, The University of Tokyo, Japan*
[4] *Environmental and Radiological Health Sciences, Colorado State University, Fort Collins, USA*

IODINE-129 IN AEROSOLS FROM GERMANY

Use of AMS for determination of monthly concentration of ^{129}I in air

A. Daraoui[1], M. Schwinger[1], B. Riebe[1], C. Walther[1], C. Vockenhuber, H.-A. Synal

Here, we report on concentrations of ^{129}I and ^{127}I, and ^{129}I/^{127}I isotopic ratios in aerosols over Germany. Aerosol filter samples collected in 2012 by the German Meteorological Service (DWD) and the Federal Office for Radiation Protection (BfS) were analyzed with ICP-MS and AMS after aqueous extraction of iodine.

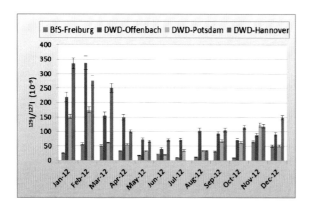

Fig. 1: ^{127}I concentration vs. ^{129}I concentration.

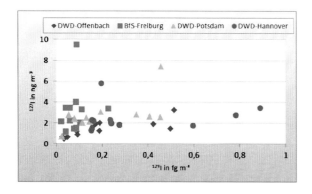

Fig. 2: Time profile of ^{129}I/^{127}I isotopic ratios.

The mean of the ^{127}I values varied between 1.4 and 3.1 ng/m³. No clear correlation between ^{127}I and ^{129}I is observed (Fig. 1) suggesting that ^{127}I and ^{129}I have different sources. The dominant source for both ^{127}I and ^{129}I is the ocean, while gaseous emissions from reprocessing plants are a supplementary source for ^{129}I. ^{129}I values ranged from 0.02 to 0.89 fg/m³. Our values are

in the same range as those from Sweden (0.01 - 0.87 fg/m³) [1], but higher than results from Austria (0.02 - 0.20 fg/m³) [2].

The ^{129}I/^{127}I ratios ranged between 0.3×10^{-7} and 1.4×10^{-7} (Fig. 2). Generally, the highest ratios were found in Northern and Western Germany. A similar pattern can be seen for ^{129}I. The temporal variation of ^{129}I seems to follow the release from Sellafield except for station Freiburg that is more in line with the release pattern of La Hague (Fig. 3). However, these results cannot yet conclusively explain, whether liquid or gaseous emissions dominate (^{129}I concentrations in aerosols).

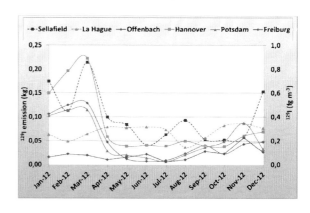

Fig. 3: ^{129}I concentrations in aerosols (right axis, solid lines), and in monthly gaseous emission from reprocessing plants (left axis, dashed lines).

[1] E. Englund et al., Nucl. Instr. & Meth. B 268 (2010) 1139

[2] T. Jabbar et al., Nucl. Instr. & Meth. B 269 (2011) 3183

---

[1] Radioecology and Radiation Protection, University of Hannover, Germany

^{129}I IN EXHAUST AIR OF GERMAN NUCLEAR FACILITIES

Expanding the sensitivity limits of γ-spectrometry with AMS

C. Brummer[1], A. Heckel[1], M. Christl, C. Vockenhuber, C. Strobl[1]

In order to validate the reliability and quality of the self-monitoring of German nuclear installations, the Federal Office for Radiation Protection (BfS) is, since more than 25 years, legally obligated by the licensing authorities to maintain a control measurement program. The legal basis for the program is the federal guideline "Verification of the Licensee's Monitoring of Radioactive Effluents from Nuclear Power Plants". According to this guideline, samples relevant for the assessment of emissions from a nuclear installation have to be sent to BfS (Fig. 1). At BfS, individual samples are selected and measured. The results are compared to the data of the facility operators. In addition, the emission of airborne particulate and gaseous ^{129}I of radioactive waste repositories and of research institutes is monitored.

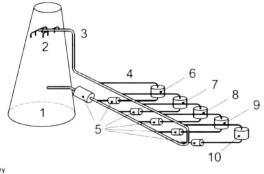

Key

1	stack	6	sampler of aerosol particles
2	rake	7	monitor for particle bound activity
3	primary sampling line	8	monitor for radioactive iodine
4	secondary sampling line	9	noble gas monitor
5	air pump	10	sampler for gaseous ^{3}H and ^{14}C

Fig. 1: *Scheme of a sampling system in a nuclear installation.*

The activity concentration of ^{129}I is determined with γ-spectrometry at BfS. The radiation of ^{129}I adsorbed on activated charcoal beds and HEPA filters (Fig. 2) is measured with a planar, high purity germanium detector (n-type) and with a

BeGe-detector (p-type). ^{129}I is identified through the k_α X-ray line of the daughter nuclide ^{129}Xe at the energy of 29.6 keV or by the γ absorption line at 39.6 keV. The detection of the X-ray line has a significantly lower detection limit due to a higher transition probability. However, the identification of ^{129}I is not unique, as several other nuclides emit at this energy.

In 2013, a charcoal sample from a prominent German nuclear facility was analyzed for the characteristic X-ray lines of ^{129}I. However, the calculated activity concentration of ^{129}I was at the detection limit determined for the 39.6 keV γ absorption line of ^{129}I. A clear identification of the radionuclide ^{129}I was thus not possible.

Fig. 2: *HEPA filter used to determine airborne, particulate ^{129}I.*

To resolve the problem, samples were measured at the ETH Zurich low-energy AMS system TANDY. The results showed clear evidence of ^{129}I in the range of the γ-spectrometric measurement of BfS.

---

[1] *Federal Office for Radiation Protection, Oberschleiß-heim, Munich, Germany*

LONG-LIVED HALOGEN NUCLIDES IN ABSORBER FOILS

^{129}I and ^{36}Cl in the absorber foils of the MEGAPIE expansion tank

B. Hammer[1,2], A. Türler[1,2], D. Schumann[2], J. Neuhausen[2], V. Boutellier[3], M. Wohlmuther[4], C. Vockenhuber

The long-lived halogen radionuclides ^{129}I and ^{36}Cl are produced, among others, in the PSI (Paul Scherrer Institute) liquid metal spallation target MEGAPIE (Fig. 1). An absorber was installed in the expansion volume to catch these volatile species during operation. First results on the adsorption of these nuclides are presented.

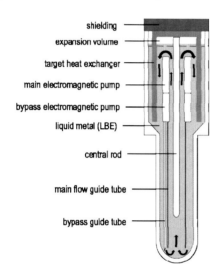

Fig. 1: *Schematic view of the MEGAPIE target.*

Silver has been selected as primary absorber material due to its capability to bind Hg as well as iodine and chlorine. A Pd foil was added to investigate its performance as absorber. The absorber consists of 6 Ag foils and one Pd foil, staggered onto each other and was mounted in the target expansion volume on the central rod (Fig. 1) in a position safely above the free surface of the liquid metal (LBE).

The absorber was retrieved from the irradiated target and small pieces (0.2 to 1 g) of each foil not contaminated by splashed LBE were cut off and used for the chemical separation [1, 2]. The ^{129}I samples were measured with the TANDY and ^{36}Cl samples measured at the 6 MV EN Tandem.

Table 1 summarizes the presently available results for ^{129}I and ^{36}Cl. Compared to FLUKA and MCNPX calculations [3], the evaporated ^{129}I and ^{36}Cl amounts represent only a small fraction of the total activities.

Absorber	m [g]	^{129}I [Bq]	^{36}Cl [Bq]
1 (Ag)	9.6299	1.81E-09	2.65
2 (Ag)	9.6875	3.47E-05	2.93
3 (Ag)	9.3945	1.00E-05	1.08
4 (Ag)	9.3988	1.49E-05	9.16
5 (Ag)	9.8407	8.46E-06	1.23
6 (Pd)	2.3433	1.36E-05	0.09
7 (Ag)	9.6419	1.19E-05	-

Tab. 1: *Activities of ^{129}I and ^{36}Cl found on the Ag and Pd foils (end of beam: 21.12.2006).*

These first preliminary studies demonstrate that Pd and Ag may serve as suitable catchers for volatiles in a possible future facility like MYRRHA [4]. For a serious evaluation of the feasibility, however, dedicated adsorption experiments are necessary to quantify the capacity of the selected materials.

The work was funded by the EC projects ANDES and GETMAT in the EURATOM FP7 framework.

[1] B. Hammer et al., Laboratory for Radio- and Environmental Chemistry (PSI) Annual Report (2013)
[2] D. Schumann et al., Laboratory for Radio- and Environmental Chemistry (PSI) Annual Report (2007) 43
[3] L. Zanini et al., PSI Report nr. 08-04 (2008)
[4] http://myrrha.sckcen.be/

[1] *Chemistry and Biochemistry, University of Bern*
[2] *Radio- and Environmental Chemistry, PSI, Villigen*
[3] *Hotlab, PSI, Villigen*
[4] *Large Research Facilities (GFA), PSI, Villigen*

MATERIALS SCIENCES

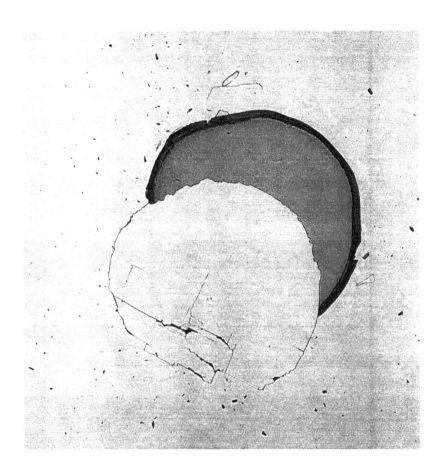

In-air microbeams: Evaluation of beam quality

Heavy ion STIM imaging

The MeV SIMS project

Al-Si separation by carbon backscattering

Oxygen transport in noble metals

Accurate RBS analysis of angular distributions

Ion beam induced selective grain growth

IN-AIR MICROBEAMS: EVALUATION OF BEAM QUALITY

Expanding the applicability of the capillary microprobe to heavy ions

M. Schulte-Borchers, A. Eggenberger, M. Döbeli, A.M. Müller

The capillary microprobe is a well suited tool to extract high quality microbeams of light ions into air. Its capability to collimate even heavy particles motivated us to investigate the extraction of heavy ion microbeams into air, which would be of benefit to specific in-air applications (like Scanning Transmission Ion Microscopy, STIM). To evaluate the beam quality generated this way nuclear track detectors were irradiated. The positions of developed single ion tracks were digitized using a Mathematica® routine (Fig. 1).

The lateral distribution of ion hits proved that the beam spread of heavy ion beams in air is larger than for light ions. The residual air inside the capillary leads to energy loss and angular straggling, which is higher for heavy ions.

Evaluation of energy spectra showed a degeneration of the full-energy peak. The energy loss can however be reduced by using a He gas atmosphere instead of air (Fig.2). The full-energy peak is also narrower.

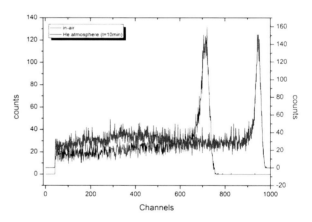

Fig. 2: *Energy spectra of a 2 MeV p beam in air (black) and in a He gas atmosphere (red).*

Our experiments showed that it is possible with the capillary microprobe to extract heavy ion microbeams into air. As expected, beam quality suffers more compared to light ions. Nevertheless, sufficiently narrow full-energy peaks and beam diameters in the micrometer range were achieved. The possibility to use a light gas atmosphere instead of air to limit energy loss and straggling is promising for experiments, where beam energy and diameter are more critical.

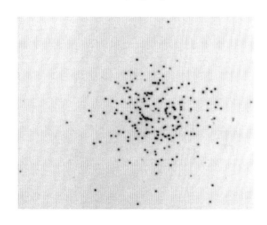

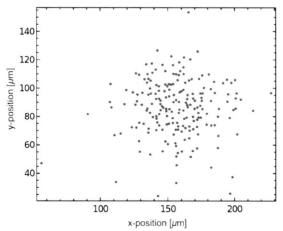

Fig. 1: *Microscope image and digitized positions of single ion hits in a track detector.*

HEAVY ION STIM IMAGING

Comparing imaging contrast of different ions on a drosophila wing

M. Schulte-Borchers, M. Döbeli, A.M. Müller, M.J. Simon

STIM (Scanning Transmission Ion Microscopy) is a well-understood procedure to image density distributions in thin samples. By measuring the number and energy of transmitted particles through the sample, conclusions can be drawn on the density of e.g. biological samples.

The capillary microprobe enables STIM measurements in air without restrictions on projectile masses, thereby enabling Heavy Ion STIM. The increased energy loss of heavy ions compared to proton or helium projectiles is expected to induce higher contrast in the resulting images.

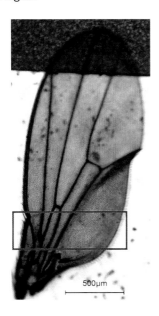

Fig. 1: *Microscopic image of the drosophila wing.*

As an example for a thin specimen, a section of a drosophila wing (Fig. 1, marked in red) was imaged with STIM using 1 MeV protons and 13 MeV iodine ions. A glass capillary with 0.7 μm outlet diameter was used for the microbeam extraction into air. Ion detection was made with a gas ionization detector.

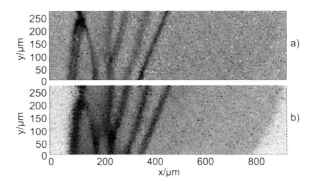

Fig. 2: *STIM images of the same sample area measured with 1 MeV p (a) and 13 MeV I (b).*

The images obtained (Fig. 2) indicate higher contrast in the iodine measurement. The structure of longitudinal veins is more pronounced and the transition from wing skin to air can be clearly observed (area in the lower right corner) in contrast to the proton image.

Line scans have been made to quantify the differences. The iodine curve verifies the rapid increase in contrast and the higher level of detail (Fig. 3).

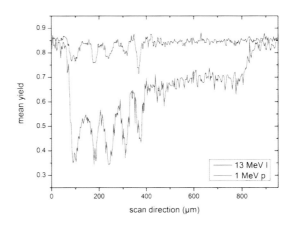

Fig. 3: *Line scans (at y = 25 μm) from both images show the strong enhancement of contrast in the iodine measurement (red line).*

THE MEV SIMS PROJECT

Designing an MeV SIMS setup based on a capillary microprobe

M. Schulte-Borchers, M. Döbeli, M. Klöckner, A.M. Müller, H.-A. Synal

Secondary Ion Mass Spectrometry (SIMS) with MeV primary ion beams has the advantage of higher molecular secondary ion yields compared to traditional keV SIMS [1]. This is of benefit to those medical and biological applications where molecular imaging is often needed to understand the underlying principles. An increase in molecular secondary ion yields has also been achieved with heavy ions or cluster particles in keV SIMS [2].

With a capillary microprobe both approaches could be combined, because heavy ions and clusters can be collimated to micrometer diameters. This motivated us to investigate molecular ion yields with heavy and cluster MeV ions.

A new experimental setup is currently in the planning phase. It will employ a goniometer and rotator combination (see Fig. 1) for capillary angle adjustment. Earlier experiments showed that adjustments with a precision of $\approx 0.25°$ over a total range of 2° is sufficient, which is feasible with commercially available nano-positioners. Due to varying capillary lengths also the distance of the setup to the sample holder position has to be adjustable.

As the capillary microbeam cannot be scanned across the sample by electrostatic deflection, the sample stage will be equipped with a piezo scanning stage to move the sample instead.

Current developments involve a sample changing system for three to five samples to be loaded into vacuum at a time.

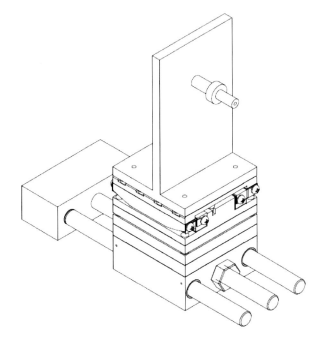

Fig. 1: *Schematic of the rotator-goniometer stage for capillary angle adjustment with capillary holder mounted on top.*

In addition, a simple linear time-of-flight spectrometer is being designed based on SIMION simulations. It involves two-field ion acceleration and a flight path of 0.5 m. At its end an MCP detector will detect the atomic and molecular ions from the sample surface. All variable parameters (extraction voltages, grid sizes, distance) will be optimized in simulations for optimal mass resolution.

[1]   B.N. Jones et al., Surface and Interface Analysis 43 (1-2) (2011) 249

[2]   J.S. Fletcher et al., Analytical Chemistry 78 (6) (2006) 1827

Al-Si SEPARATION BY CARBON BACKSCATTERING

Solving the interference in RBS spectra of Al films on Si

M. Döbeli, A.M. Müller, J. Ramm[1], A. Dommann[2], X. Mäder[3], A. Neels[3], D. Passerone[4], D. Scopece[4]

It is inherently difficult to measure thin layers of Al on Si substrates by He RBS. The Si edge is 37 keV higher than the one of Al but the energy loss of the backscattered He particles in an Al film pushes the Si edge back so that there is always an interference peak between the signals of the two elements almost independent of the layer thickness (Fig. 1).

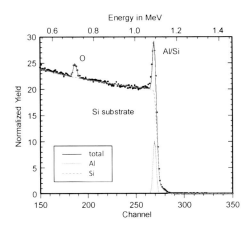

Fig. 1: *2 MeV He RBS spectra of a partially oxidized 11 nm thick Al film on Si.*

Heavy Ion Backscattering Spectrometry (HIBS) with 10 MeV ^{12}C ions [1] greatly improves the situation (Fig. 2).

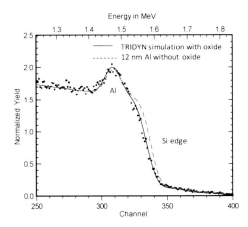

Fig. 2: *10 MeV ^{12}C HIBS spectrum of the same sample as in Fig. 1.*

With the enhanced mass and depth resolution of HIBS in combination with a high resolution gas ionization detector the Al signal clearly sits on the Si background with no interference peak.

It is even possible to resolve the position of an oxidized layer below the Al film deposited by a cathodic arc process on the Si substrate (Fig. 2). This allows now to verify results of TRIDYN Monte-Carlo calculations [2]. Simulated concentration depth profiles (Fig. 3) are directly fed into the RBS simulator and the sequence of layers can be identified by comparison with the HIBS data (Fig. 2). This technique can potentially be refined by a further enhancement of the gas detector resolution.

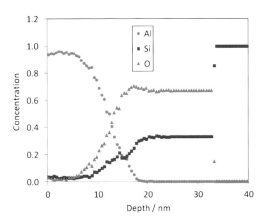

Fig. 3: *Al, Si and O depth profiles simulated for Al deposited on SiO_2/Si at an energy of 800 eV by a cathodic arc process.*

[1]   M. Mallepell et al., Nucl. Instr. & Meth. B267 (2009) 1193

[2]   W. Möller and W. Eckstein, Nucl. Instr. & Meth. B2 (1984) 814

[1] *Oerlikon Balzers Coating AG, Balzers, Liechtenstein*
[2] *Materials meet life, EMPA, St. Gallen*
[3] *Microsystems Technology, CSEM, Neuchâtel*
[4] *Nanotech@surfaces, EMPA, Dübendorf*

OXYGEN TRANSPORT IN NOBLE METALS

Ceramic-metal multilayers analyzed with APT and RBS

B. Scherrer[1, 2], H. Galinski[2, 3], J. Cairney[1], M. Döbeli

Grain boundaries and triple junctions of nanocrystalline Pt thin films have shown oxygen reduction capability [1]. In the present project the oxygen transport mechanism in this noble metal is investigated with a correlative approach of analytical techniques with outstanding resolution. This offers interesting perspectives on the development of new catalytic materials and gives new insight into the thermodynamics of nanostructured thin films.

Pt/ZrO_2 multilayers have been deposited by magnetron sputtering. The as-deposited films have been analyzed by 2 and 5 MeV He RBS to determine layer composition and thickness (Fig. 1). Thicknesses correspond well with values determined by Transmission Electron Microscopy. Samples suited for Atom Probe Tomography (APT) were then prepared by focused ion beam milling in horizontal (Fig. 2) and vertical orientation. APT allows the tomographic reconstruction of the sample structure with virtually atomic resolution. An overall view of an APT with vertical oriented multilayers is shown in Fig. 3. Oxygen excess was found at the interface between Pt and ZrO_2. This suggests that the oxygen is located at the platinum triple junctions and/or the grain boundaries.

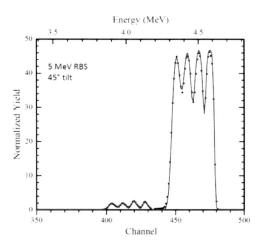

Fig. 1: *5 MeV He RBS spectrum of a Pt/ZrO_2 multilayer sample. Average thickness is (41±1) nm for Pt and (13±1) nm for ZrO_2.*

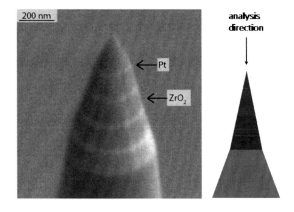

Fig. 2: *SEM micrograph of a tip prepared for APT from the layer system shown in Fig. 1.*

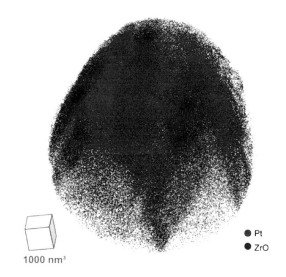

Fig. 3: *APT of a vertical oriented multilayer.*

[1]   T. Ryll et al., Phys. Rev. B84 (2011) 184111

[1] *ACMM, University of Sidney, Australia*
[2] *Nanometallurgy, ETH Zurich*
[3] *School of Eng. and Appl. Sciences, Harvard University, Boston, USA*

ACCURATE RBS ANALYSIS OF ANGULAR DISTRIBUTIONS

Scattering of laser ablated particles by a gas pulse in vacuum

P.R. Willmott [1], M.L. Reinle-Schmitt[1], C. Cancellieri [1], M. Döbeli

In Pulsed Reactive Crossed-beam Laser Ablation (PRCLA) the interaction of laser ablated particles with a short gas pulse or background gas can add to the non-thermal synthesis and growth of technologically relevant materials [1]. Scattering of the ablation plasma by the gas leads to a slight shift in composition at the deposition site.

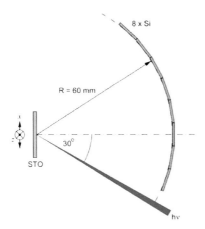

Fig. 1: *Plasma plume at the ablation target.*

The angular dependence of composition and thickness of a deposited $SrTiO_3$ (STO) film was analyzed by mounting silicon catcher substrates around the ablation target (Fig. 1 and 2).

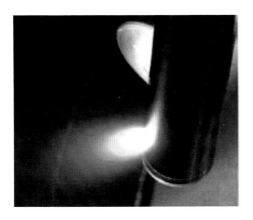

Fig. 2: *Experimental setup of silicon catcher substrates for the measurement of angular distributions.*

The films on the catcher substrates were measured by 2 MeV He RBS. Since elemental peaks in the RBS spectra are well isolated and scattering cross-sections are known to a fraction of a percent the Sr to Ti ratio can be determined with an absolute accuracy of 0.3 to 0.5 %.

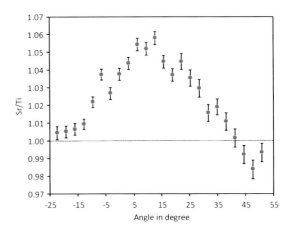

Fig. 3: *Sr to Ti ratio as a function of substrate position.*

The angular dependence of the composition of a film grown at $8 \cdot 10^{-5}$ mbar oxygen background pressure is shown in Fig. 3.

The Sr content is clearly enhanced under normal emission to the target surface. This can be consistently explained by the higher probability for the lighter Ti atoms to be scattered under larger angles by gas atoms. The dependence of the film thickness is mostly determined by the shape of the plasma plume.

[1] P.R. Willmott and J.R. Huber, Rev. Mod. Phys. 72 (2000) 315

[1] *Materials Science Group, Paul Scherrer Institut (PSI), Villigen*

ION BEAM INDUCED SELECTIVE GRAIN GROWTH

In-plane texture development in bcc refractory metallic thin films

H. Ma[1], M. Seita[2], R. Spolenak[1], M. Döbeli, M. Schulte-Borchers

Microstructure control (e.g. grain size, defect content and texture) in thin films is an effective way to achieve specific material properties. Generally, such control can be realized simply by thermal annealing at a temperature higher than 0.3 - 0.5 melting temperature of materials. For refractory materials (e.g. Ta and W), however, this temperature is too high to be effectively achieved.

In this project, we introduce an athermal method, which enables the control of grain size and in-plane texture in refractory metallic thin films by ion bombardment. Ta and W thin films, magnetron sputtered on Si wafer, are bombarded by 4.5 MeV Au ion beams at liquid nitrogen temperature. The ion beam is set to be parallel to one of the <111> channeling direction. The evolution of grain size and texture is characterized by EBSD (Electron Backscatter Diffraction).

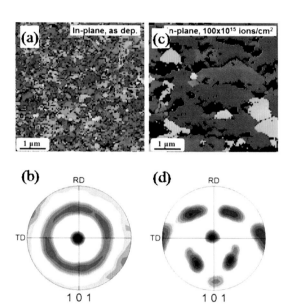

Fig. 2: *EBSD in-plane orientation maps ((a) and (c)) and (110) pole figures ((b) and (d)) of a 500 nm thick W film before ((a) and (b)) and after ((c) and (d)) Au-ion bombardment at 4.5 MeV at 77 K.*

The film microstructure before and after ion bombardment are shown in Fig. 1 and Fig. 2 for Ta [1] and W, respectively. For both cases, the as deposited film shows (110) fiber texture with small grain size. The ion bombardment led to significant grain growth and in-plane texture development. These changes can be explained by ion beam induced selective grain growth, where the subset of grains in channeling orientation grows at the expense of the others in order to minimize the volume free energy.

[1] M. Seita et al., Appl. Phys. Lett. 101 (2012) 251905

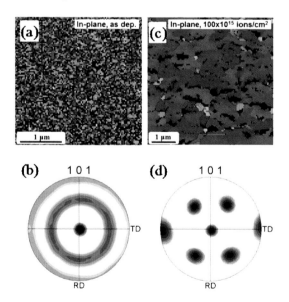

Fig. 1: *EBSD in-plane orientation maps ((a) and (c)) and (110) pole figures ((b) and (d)) of a 400 nm thick Ta film before ((a) and (b)) and after ((c) and (d)) Au-ion bombardment at 4.5 MeV at 77 K.*

[1] *Nanometallurgy, ETH Zurich*
[2] *Materials Science and Engineering, MIT, Boston, USA*

EDUCATION

Books and their ages

Presentation awards for Ph.D. students

BOOKS AND THEIR AGES

Dating paper from the 19th/20th century

I. Hajdas, L. Jakob[1], T. Richard[1], M. Maurer, L. Wacker

Old books are fascinating objects often desired by connoisseurs and therefore becoming expensive when proven old. The question of reliability when dating paper is connected to the technique of paper production, type and sources of material used. Since the industrial revolution, paper is produced from wood pulp which can show a ^{14}C content that reflects a mixture of old and young trees. As an effect of nuclear weapons testing on the radiocarbon content in trees, paper produced after 1950 AD with trees grown from before would result in ages too old. Such results were observed in a similar project that focused on dating library books from the 1950s and the 1960s.

As part of a one-week school project organized by Kantonsschule Olten, which involved participation in research at the ETH, we have chosen to radiocarbon date the paper of two books from the end of 19th and the beginning of 20th century (Fig. 1).

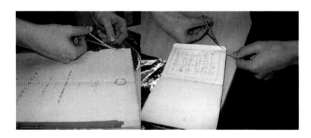

Fig. 1: *Samples taken from a geology book published in 1885 (left) and a scout book ('Pfadibüechli') with songs of Swiss scouts (Pfadfinder Lieder) from the early 20th century (date not known).*

Paper samples were treated with weak solutions of acid, base and then again acid to remove potential contamination with old and young carbon [1].

The radiocarbon dates do not result in precise calendar ages due to the fact that the atmospheric ^{14}C concentration during the period of the 17th -20th century underwent significant variation resulting in 'wiggles' in the radiocarbon calibration curve (Fig. 2).

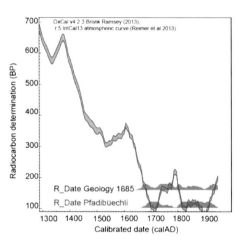

Fig. 2: *Calibration of the ^{14}C ages of the paper samples results in a wide range of calendar ages (mid 17th to mid-20th century).*

The calibrated ^{14}C ages of the paper from both books show that the material used for production of the paper had an age not older than 200 years at the time of the print. It cannot be excluded that relatively young wood was used to create the pulp (Fig. 2).

On the other hand, other type of source material (straw, flax, cotton etc.) can be used to produce paper. Thus a deeper investigation might provide interesting information on the history of paper production in Switzerland.

[1] I. Hajdas, QSJ - Eiszeitalter und Gegenwart, 57 (2008) 2

---

[1] *Kantonsschule Olten*

PRESENTATION AWARDS FOR PHD STUDENTS

Two Ph.D. students received prices for scientific presentations

Scientific staff

Two Ph.D. students from the Laboratory of Ion Beam Physics received presentation awards at scientific conferences this year.

Martina Schulte-Borchers won the second price in the Graduation Student Poster Competition during the International Conference on Ion Beam Analysis 2013 held June 23-28 in Seattle, USA [1]. The title of her poster was "Heavy ion microbeams extracted into air using a capillary microprobe". The poster focused on the effect of residual air on in-air heavy ion microbeams from the capillary microprobe and showed results from an application of such microbeams in Heavy Ion STIM imaging.

Fig. 1: Guest of honor James Ziegler presents Martina Schulte-Borchers with the second price in the Graduation Student Poster Competition at the IBA conference banquet.

At the Fall Meeting 2013 of the Swiss Chemical Society (SCS) on September 6 in Lausanne, Switzerland [1], Caroline Münsterer was honored with the Metrohm award as the runner-up in the Best Oral Presentation Competition in the Analytical Sciences section. In her presentation "^{14}C analysis of carbonates at high spatial resolution by the coupling of Laser Ablation with AMS" she showed the first

results of her Ph.D. project which investigates the CO_2 production rate from carbonates during laser ablation and ^{14}C measurements from a Laser Ablation apparatus coupled to the Zurich MICADAS AMS facility.

Fig. 2: Marcus Tobler, CEO of Metrohm Schweiz AG, presents the Metrohm Oral Presentation Award to Caroline Münsterer at the SCS Fall Meeting.

[1] http://iba2013.labworks.org
[2] http://www.scg.ch/fallmeeting

PUBLICATIONS

J.A. Abreu, J. Beer, F. Steinhilber, M. Christl and P.W. Kubik
^{10}Be in ice cores and ^{14}C in tree rings: Separation of production and climate effects
Space Science Reviews **176** (2013) 343-349

F. Adolphi, D. Güttler, L. Wacker, G. Skog and R. Muscheler
Intercomparison of C-14 dating of wood samples at Lund University and ETH-Zurich AMS facilities: Extraction, graphitization and measurement
Radiocarbon **55** (2013) 391-400

V. Alfimov, A. Aldahan and G. Possnert
Water masses and ^{129}I distribution in the Nordic Seas
Nuclear Instruments and Methods B **294** (2013) 542-546

R. Belmaker, B. Lazar, J. Beer and M. Christl
^{10}Be dating of Neogene halite
Geochimica and Cosmochimica Acta **122** (2013) 418-429

A. Birkholz, R.H. Smittenberg, I. Hajdas, L. Wacker and S.M. Bernasconi
Isolation and compound specific radiocarbon dating of terrigenous branched glycerol dialkyl glycerol tetraethers (brGDGTs)
Organic Geochemistry **60** (2013) 9-19

D.K. Bora, A. Braun, R. Erni, U. Müller, M. Döbeli and E.C. Constable
Hematite–NiO/a-Ni(OH)$_2$ heterostructure photoanodes with high electrocatalytic current density and charge storage capacity
Physical Chemistry Chemical Physics **15** (2013) 12648-12659

L. Calcagnile, G. Quarta, L. Maruccio, H.-A. Synal and A.M. Müller
Design features of the new multi isotope AMS beamline at CEDAD
Nuclear Instruments and Methods B **294** (2013) 416-419

F. Cao, Y.L. Zhang, S. Szidat, A. Zapf, L. Wacker and M. Schwikowski
Microgram-level radiocarbon determination of carbonaceous particles in firn and ice samples: Pretreatment and OC/EC separation
Radiocarbon **55** (2013) 383-390

M. Christl and P.W. Kubik
New Be-cathode preparation method for the ETH 6 MV Tandem
Nuclear Instruments and Methods B **294** (2013) 199-202

M. Christl, J. Lachner, C. Vockenhuber, I. Goroncy, J. Herrmann and H.-A. Synal
First data of Uranium-236 in the North Sea
Nuclear Instruments and Methods B **294** (2013) 530-536

M. Christl, C. Vockenhuber, P.W. Kubik, L. Wacker, J. Lachner, V. Alfimov and H.-A. Synal
The ETH Zurich AMS facilities: Performance parameters and reference materials
Nuclear Instruments and Methods B **294** (2013) 29-38

S.M. Fahrni, L. Wacker, H.-A. Synal and S. Szidat
Improving a gas ion source for ^{14}C AMS
Nuclear Instruments and Methods B **294** (2013) 320-327

J. Fallis, A. Parikh, P.F. Bertone, S. Bishop, L. Buchmann, A.A. Chen, G. Christian, J.A. Clark, J.M. D'Auria, B. Davids, C.M. Deibel, B.R. Fulton, U. Greife, B. Guo, U. Hager, C. Herlitzius, D.A. Hutcheon, J. Jos, A.M. Laird, E.T. Li, Z.H. Li, G. Lian, W.P. Liu, L. Martin, K. Nelson, D. Ottewell, P.D. Parker, S. Reeve, A. Rojas, C. Ruiz, K. Setoodehnia, S. Sjue, C. Vockenhuber, B. Wang and C. Wrede
Constraining nova observables: Direct measurements of resonance strengths in $^{33}S(p,\gamma)^{34}Cl$
Physical Review C **88** (2013)

X.J. Feng, J.E. Vonk, B.E. van Dongen, O. Gustafsson, I.P. Semiletov, O.V. Dudarev, Z.H. Wang, D.B. Montlucon, L. Wacker and T.I. Eglinton
Differential mobilization of terrestrial carbon pools in Eurasian Arctic river basins
Proceedings of the National Academy of Sciences of the United States of America **110** (2013) 14168-14173

R. Frison, S. Heiroth, J.L.M. Rupp, K. Conder, E.J. Barthazy, E. Müller, M. Horisberger, M. Döbeli and L.J. Gauckler
Crystallization of 8 mol% yttria-stabilized zirconia thin-films deposited by RF-sputtering
Solid State Ionics **232** (2013) 29-36

E.F. Gjermundsen, J.P. Briner, N. Akçar, O. Salvigsen, P.W. Kubik, N. Gantert and A. Hormes
Late Weichselian local ice dome configuration and chronology in Northwestern Svalbard: early thinning, late retreat
Quaternary Science Reviews **72** (2013) 112-127

K. Guelland, F. Hagedorn, R.H. Smittenberg, H. Göransson, S.M. Bernasconi, I. Hajdas and R. Kretzschmar
Evolution of carbon fluxes during initial soil formation along the forefield of Damma glacier, Switzerland
Biogeochemistry **113** (2013) 545-561

D. Güttler, L. Wacker, B. Kromer, M. Friedrich and H.-A. Synal
Evidence of 11-year solar cycles in tree rings from 1010 to 1110 AD – Progress on high precision AMS measurements
Nuclear Instruments and Method B **294** (2013) 459-463

R.L. Hermanns, T. Oppikofer, H. Dahle, T. Eiken, S. Ivy-Ochs and L.H. Blikra
Understanding long-term slope deformation for stability assessment of rock slopes: the case of the Oppstadhornet rockslide, Norway
Engineering Geology and Environment **6** (2013) 255-264

K. Hippe, F. Kober, L. Wacker, S.M. Fahrni, S. Ivy-Ochs, N. Akçar, C. Schlüchter and R. Wieler
An update on in situ cosmogenic ^{14}C analysis at ETH Zürich
Nuclear Instruments and Methods B **294** (2013) 81-86

A. Hogg, C. Turney, J. Palmer, J. Southon, B. Kromer, C.B. Ramsey, G. Boswijk, P. Fenwick, A. Noronha, R. Staff, M. Friedrich, L. Reynard, D. Guetter, L. Wacker and R. Jones
The New Zealand kauri (Agathis australis) research project: A radiocarbon dating intercomparison of Younger Dryas wood and implications for IntCal13
Radiocarbon **55** (2013) 2035-2048

S. Ivy-Ochs, M. Dühnforth, A.L. Densmore and V. Alfimov
Dating fan deposits with cosmogenic nuclides
Global Change Research **47** (2013) 243-263

R. Janovics, Z. Kern, D. Güttler, L. Wacker, I. Barnabas and M. Molnar
Radiocarbon impact on a nearby tree of a light-water VVER-type nuclear power plant, Paks, Hungary
Radiocarbon **55** (2013) 826-832

T.D. Jones, I.T. Lawson, J.M. Reed, G.P. Wilson, M.J. Leng, M. Gierga, S.M. Bernasconi, R.H. Smittenberg, I. Hajdas, C.L. Bryant and P.C. Tzedakis
Diatom-inferred late Pleistocene and Holocene palaeolimnological changes in the Ioannina basin, northwest Greece
Journal of Paleolimnology **49** (2013) 185-204

D.J. Kennett, I. Hajdas, B.J. Culleton, S. Belmecheri, S. Martin, H. Neff, J. Awe, H.V. Graham, K.H. Freeman, L. Newsom, D.L. Lentz, F.S. Anselmetti, M. Robinson, N. Marwan, J. Southon, D.A. Hodell and G.H. Haug
Correlating the ancient Maya and modern European calendars with high-precision AMS ^{14}C dating
Scientific Reports **3, 1597** (2013) DOI: 10.1038/srep01597

F. Kober, G. Zeilinger, S. Ivy-Ochs, A. Dolati, J. Smit and P.W. Kubik
Climatic and tectonic control on fluvial and alluvial fan sequence formation in the Central Makran Range, SE-Iran
Global and Planetary Change **111** (2013) 133-149

B. Kromer, S. Lindauer, H.-A. Synal and L. Wacker
MAMS – A new AMS facility at the Curt-Engelhorn-Centre for Achaeometry, Mannheim, Germany
Nuclear Instruments and Methods B **294** (2013) 11-13

M.S. Krzemnicki and I. Hajdas
Age determination of pearls: A new approach for pearl testing and identification
Radiocarbon **55** (2013) 1801-1809

J. Kuhlemann, E. Gachev, A. Gikov, S. Nedkov, I. Krumrei and P.W. Kubik
Glaciation in the Rila mountains (Bulgaria) during the Last Glacial Maximum
Quaternary International **293** (2013) 51-62

J. Lachner, M. Christl, H.-A. Synal, M. Frank and M. Jakobsson
Carrier free $^{10}Be/^9Be$ measurements with low-energy AMS: Determination of sedimentation rates in the Arctic Ocean
Nuclear Instruments and Methods B **294** (2013) 67-71

J. Lachner, M. Christl, C. Vockenhuber and H.-A. Synal
Detection of UH^{3+} and ThH^{3+} molecules and ^{236}U background studies with low-energy AMS
Nuclear Instruments and Methods B **294** (2013) 364-368

S.Q. Lang, G.L. Früh-Green, S.M. Bernasconi and L. Wacker
Isotopic ($\delta^{13}C$, $\Delta^{14}C$) analysis of organic acids in marine samples using wet chemical oxidation
Limnology and Oceanography-Methods **11** (2013) 161-175

F. Lavigne, J.-P. Degeai, J.-C. Komorowski, S. Guillet, Vincent Robert, P. Lahitte, C. Oppenheimer, M. Stoffel, C.M. Vidal, Surono, I. Pratomo, P. Wassmer, I. Hajdas, D.S. Hadmoko and E. de Belizal
Source of the great A.D. 1257 mystery eruption unveiled, Samalas volcano, Rinjani Volcanic Complex, Indonesia
Proceedings of the National Academy of Sciences of the United States of America **110** (2013) 16742-16747

J. Leifeld, S. Bassin, F. Conen, I. Hajdas, M. Egli and J. Fuhrer
Control of soil pH on turnover of belowground organic matter in subalpine grassland
Biogeochemistry **112** (2013) 59-69

M. Macchia, M. D'Elia, G. Quarta, V. Gaballo, E. Braione, L. Maruccio, L. Calcagnile, G. Ciceri, V. Martinotti and L. Wacker
Extraction of dissolved inorganic carbon (DIC) from seawater samples at CEDAD: Results of an intercomparison
Radiocarbon **55** (2013) 579-584

M. Martschini, P. Andersson, O. Forstner, R. Golser, D. Hanstorp, A.O. Lindahl, W. Kutschera, S. Pavetich, A. Priller, J. Rohlén, P. Steier, M. Suter and A. Wallner
AMS of ^{36}Cl with the VERA 3 MV tandem accelerator
Nuclear Instruments and Methods B **294** (2013) 115-120

M. Molnár, I. Hajdas, R. Janovics, L. Rinyu, H.-A. Synal, M. Veres and L. Wacker
C-14 analysis of groundwater down to the millilitre level
Nuclear Instruments and Methods B **294** (2013) 573 - 576

M. Molnár, R. Janovics, I. Major, J. Orsovszki, R. Gonczi, M. Veres, A.G. Leonard, S.M. Castle, T.E. Lange, L. Wacker, I. Hajdas and A.J.T. Jull
Status report of the new AMS C-14 sample preparation lab of the Hertelendi Laboratory of Environmental Studies (Debrecen, Hungary)
Radiocarbon **55** (2013) 665-676

M. Molnár, L. Rinyu, M. Veres, M. Seiler, L. Wacker and H.A. Synal
Environmicadas: A Mini C-14 AMS with Enhanced Gas Ion Source Interface in the Hertelendi Laboratory of Environmental Studies (Hekal), Hungary
Radiocarbon **55** (2013) 338-344

A.M. Müller, M. Suter, D. Fu, X. Ding, K. Liu, H.-A. Synal and L. Zhou
A simple way to upgrade a compact radiocarbon AMS facility for ^{10}Be
Radiocarbon **55** (2013) 231-236

E. Nottoli, M. Arnold, G. Aumaître, D.L. Bourlès, K. Keddadouche and M. Suter
The physics behind the isobar separation of ^{36}Cl and ^{10}Be at the French AMS national facility ASTER
Nuclear Instruments and Method B **294** (2013) 397-402

J. Qiao, X. Hou, P. Roos, J. Lachner, M. Christl and Y. Xu
Sequential injection approach for simultaneous determination of ultratrace plutonium and neptunium in urine with Accelerator Mass Spectrometry
Analytical Chemistry **85** (2013) 8826-8833

P.J. Reimer, E. Bard, A. Bayliss, J.W. Beck, P.G. Blackwell, C.B. Ramsey, D.M. Brown, C.E. Buck, R.L. Edwards, M. Friedrich, P.M. Grootes, T.P. Guilderson, H. Haflidason, I. Hajdas, C. Hatté, T.J. Heaton, A.G. Hogg, K.A. Hughen, K.F. Kaiser, B. Kromer, S.W. Manning, R.W. Reimer, D.A. Richards, E.M. Scott, J.R. Southon, C.S.M. Turney and J. van der Plicht
Selection and treatment of data for radiocarbon calibration: an update to the international calibration (IntCal) criteria
Radiocarbon **55** (2013) 1923-1945

P.J. Reimer, E. Bard, A. Bayliss, J.W. Beck, P.G. Blackwell, C.B. Ramsey, C.E. Buck, H. Cheng, R.L. Edwards, M. Friedrich, P.M. Grootes, T.P. Guilderson, H. Haflidason, I. Hajdas, C. Hatté, T.J. Heaton, D.L. Hoffmann, A.G. Hogg, K.A. Hughen, K.F. Kaiser, B. Kromer, S.W. Manning, M. Niu, R.W. Reimer, D.A. Richards, E.M. Scott, J.R. Southon, R.A. Staff, C.S.M. Turney and J. van der Plicht
IntCal13 and Marine13 radiocarbon age calibration curves 0–50,000 years cal BP
Radiocarbon **55** (2013) 1869-1887

J. Rethemeyer, R.-H. Fülöp, S. Höfle, L. Wacker, S. Heinze, I. Hajdas, U. Patt, S. König, B. Stapper and A. Dewald
Status report on sample preparation facilities for ^{14}C analysis at the new CologneAMS center
Nuclear Instruments and Methods B **294** (2013) 168-172

A.K. Richter, I. Hajdas, E. Frossard and I. Brunner
Plant Biosystems - An international journal dealing with all aspects of plant biology: Official journal of the Societa Botanica Italiana
Plant Biosystems **147** (2013) 50-59

L. Rinyu, M. Molnár, I. Major, T. Nagy, M. Veres, Á. Kimák, L. Wacker and H.-A. Synal
Optimization of sealed tube graphitization method for environmental C-14 studies using MICADAS
Nuclear Instruments and Methods B **294** (2013) 270-275

T.E. Scharf, A.T. Codilean, M.d. Wit, J.D. Jansen and P.W. Kubik
Strong rocks sustain ancient postorogenic topography in southern Africa
Geology **41** (2013) 331-334

M.V.F. Schlupp, A. Kurlov, J. Hwang, Z. Yáng, M. Döbeli, J. Martynczuk, M. Prestat, J.-W. Son and L.J. Gauckler
Gadolinia doped ceria thin films prepared by aerosol assisted chemical vapor deposition and applications in intermediate-temperature solid oxide fuel cells
Fuel Cells **13** (2013) 658-665

S. Schneider, C. Walther, S. Bister, V. Schauer, M. Christl, H.-A. Synal, K. Shozugawa and G. Steinhauser
Plutonium release from Fukushima Daiichi fosters the need for more detailed investigations
Scientific Reports **3, 2988** (2013) DOI: 10.1038/srep02988

F. Seibert, M. Döbeli, D.M. Fopp-Spori, K. Glaentz, H. Rudigier, N. Schwarzer, B. Widrig and J. Ramma
Comparison of arc evaporated Mo-based coatings versus Cr_1N_1 and ta–C coatings by reciprocating wear test
Wear **298-299** (2013) 14-22

M. Seita, R. Schäublin, M. Döbeli and R. Spolenak
Selective ion-induced grain growth: Thermal spike modeling and its experimental validation
Acta Materialia **61** (2013) 6171-6177

D. Stender, S. Cook, J.A. Kilner, M. Döbeli, K. Conder, T. Lippert and A. Wokaun
SIMS of thin films grown by pulsed laser deposition on isotopically labeled substrates
Solid State Ionics **249-250** (2013) 56-62

H.-A. Synal
Developments in accelerator mass spectrometry
International Journal of Mass Spectrometry **349-350** (2013) 192-202

H.-A. Synal, T. Schulze-König, M. Seiler, M. Suter and L. Wacker
Mass spectrometric detection of radiocarbon for dating applications
Nuclear Instruments and Methods B **294** (2013) 349-352

S. Szidat, G. Bench, V. Bernardoni, G. Calzolai, C.I. Czimczik, L. Derendorp, U. Dusek, K. Elder, M.E. Fedi,
J. Genberg, O. Gustafsson, E. Kirillova, M. Kondo, A.P. McNichol, N. Perron, G.M. Santos, K. Stenstrom,
E. Swietlicki, M. Uchida, R. Vecchi, L. Wacker, Y.L. Zhang and A.S.H. Prevot
Intercomparison of C-14 analysis of carbonaceous aerosols: Exercise 2009
Radiocarbon **55** (2013) 1496-1509

P. Thiel, J. Eilertsen, S. Populoh, G. Saucke, M. Döbeli, A. Shkabko, L. Sagarna, L. Karvonen and
A. Weidenkaff
Influence of tungsten substitution and oxygen deficiency on the thermoelectric properties of $CaMnO_{3-\delta}$
Journal of Applied Physics **114** (2013) 243707

I.G. Usoskin, B. Kromer, F. Ludlow, J. Beer, M. Friedrich, G.A. Kovaltsov, S.K. Solanki and L. Wacker
The AD775 cosmic event revisited: the Sun is to blame
Astronomy & Astrophysics **552** (2013) L3

C. Vockenhuber, V. Alfimov, M. Christl, J. Lachner, T. Schulze-König, M. Suter and H.-A. Synal
The potential of He stripping in heavy ion AMS
Nuclear Instruments and Methods B **294** (2013) 382-386

C. Vockenhuber, J. Jensen, J. Julin, H. Kettunen, M. Laitinen, Mikko Rossi, T. Sajavaara, O. Osmani,
A. Schinner, P. Sigmund and H.J. Whitlow
Energy-loss straggling of 2–10 MeV/u Kr ions in gases
The European Physical Journal D **67** (2013) 1-9

L. Wacker, S.M. Fahrni, I. Hajdas, M. Molnar, H.-A. Synal, S. Szidat and Y.L. Zhang
A versatile gas interface for routine radiocarbon analysis with a gas ion source
Nuclear Instruments and Methods B **294** (2013) 315-319

L. Wacker, R.-H. Fülöp, I. Hajdas, M. Molnár and J. Rethemeyer
A novel approach to process carbonate samples for radiocarbon measurements with helium carrier gas
Nuclear Instruments and Methods B **294** (2013) 214-217

L. Wacker, J. Lippold, M. Molnár and H. Schulz
Towards radiocarbon dating of single foraminifera with a gas ion source
Nuclear Instruments and Methods B **294** (2013) 307-310

L. Wacker, C. Münsterer, B. Hattendorf, M. Christl, D. Günther and H.-A. Synal
Direct coupling of a laser ablation cell to an AMS
Nuclear Instruments and Methods B **294** (2013) 287-290

J.S. White, M. Bator, Y. Hu, H. Luetkens, J. Stahn, S. Capelli, S. Das, M. Döbeli, T. Lippert, V.K. Malik,
J. Martynczuk, A. Wokaun, M. Kenzelmann, C. Niedermayer and C.W. Schneider
Strain-induced ferromagnetism in antiferromagnetic LuMnO$_3$ thin films
Physical Review Letters **111** (2013) 037201

J. Xu, B. Pan, T. Liu, I. Hajdas, B. Zhao, H. Yu, R. Liu and P. Zhao
Climatic impact of the Millennium eruption of Changbaishan volcano in China: New insights from high-precision radiocarbon wiggle-match dating
Geophysical Research Letters **40** (2013) 54-59

A. Zapf, A. Nesje, S. Szidat, L. Wacker and M. Schwikowski
C-14 measurements of ice samples from the Juvfonne ice tunnel, Jotunheimen, Southern Norway-Validation of a C-14 dating technique for glacier ice
Radiocarbon **55** (2013) 571-578

Y.L. Zhang, P. Zotter, N. Perron, A.S.H. Prevot, L. Wacker and S. Szidat
Fossil and non-fossil sources of different carbonaceous fractions in fine and coarse particles by radiocarbon measurement
Radiocarbon **55** (2013) 1510-1520

C.L. Zhou, M. Simon, T. Ikeda, S. Guillous, W. Iskandar, A. Méry, J. Rangama, H. Lebius, A. Benyagoub,
C. Grygiel, A. Müller, M. Döbeli, J.A. Tanis and A. Cassimi
Transmission of slow highly charged ions through glass capillaries: Role of the capillary shape
Physical Review **88** (2013) 050901

B. Zollinger, C. Alewell, C. Kneisel, K. Meusburger, H. Gartner, D. Brandova, S. Ivy-Ochs, M.W.I. Schmidt and M. Egli
Effect of permafrost on the formation of soil organic carbon pools and their physical-chemical properties in the Eastern Swiss Alps
Catena **110** (2013) 70-85

TALKS AND POSTERS

N. Akçar, D. Tikhomirov, S. Ivy-Ochs, A. Graf, F. Schlunnegger, R. Reber, A. Claude, P.W. Kubik, C. Vockenhuber, I. Hajdas and C. Schlüchter
The Valais Glacier: its disappearance from the Alpine Foreland
Switzerland, Lausanne, 15.-26.11.2013, 11th Swiss Geoscience Meeting

N. Akçar, P. Deline, S. Ivy-Ochs, P.W. Kubik and C. Schlüchter
Surface exposure dating of rock avalanche deposits in the Ferret Valley (Mont Blanc massif, Italy)
France, Paris, 27.-31.08.2013, 8th IAG International Conference on Geomorphology

C. Bayrakdar, T. Görüm, M. Durmus, S. Ivy-Ochs and N. Akçar
Reconstruction of the evolution and chronology of the Akdag rockslide (SW Turkey)
Switzerland, Lausanne, 15.-26.11.2013, 11th Swiss Geoscience Meeting

N. Bellin, V. Vanacker and P.W. Kubik
Linking topographic indices and rock uplift rates to denudation in a low uplift rate setting: Betic Cordillera, SE Spain
France, Paris, 27.-31.08.2013, 8th IAG International Conference on Geomorphology

M. Bichler, M. Reindl, H. Häusler, S. Ivy-Ochs, C. Wirsig and J. Reitner
What happened between 14 and 10 ka in the central part of the European Eastern Alps?
Austria, Vienna, 07.-12.04.2013, European Geosciences Union General Assembly 2013

J.-D. Champagnac, R. Delunel, P.W. Kubik, A.-S. Meriauxand F. Beaud
Postglacial erosion rates from the Western Alps inferred from cosmogenic nuclides measurements
Austria, Vienna, 07.-12.04.2013, European Geosciences Union Gerneral Assembly 2013

M. Christl, J. Lachner, N. Casacuberta, P. Masque, X. Dai, S. Kramer-Tremblay and H.-A. Synal
Detecting ultra-trace levels of actinides with a compact AMS system
Canada, Quebec City, 27.05.2013, 96th Canadian Chemistry Conference

M. Christl
Actinide analyses with compact, low energy Accelerator Mass Spectrometry
Canada, Chalk River, 24.05.2013, Chalk River Labs Special Seminar

M. Christl, J. Lachner, X. Dai, X. Hou and H.-A. Synal
Expanding the actinide capabilities of the compact (0.5 MV) ETH Zurich AMS systemTandy
Germany, Hannover, 22.03.2013, DPG Frühjahrstagung

M. Christl, J. Lachner, X. Dai, S. Kramer-Tremblay and H.-A. Synal
Low energy AMS of the higher actinides (Am, Cm)
Belgium, Namur, 12.09.2013, ECAART 11 Conference

M. Christl
Aktinidenmessungen im Ultra-Spurenstoff Bereich – einige Anwendungsbeispiele
Switzerland, Spiez, 18.-19.09.2013, Radiochemieseminar

A. Claude, N. Akçar, S. Ivy-Ochs, H.R. Graf, P.W. Kubik, C. Vockenhuber, A. Dehnert, M. Rahn, P. Rentzel, C. Pümpin, and C. Schlüchter
The challenge of dating Swiss Deckenschotter with cosmogenic nuclides
Switzerland, Lausanne, 15.-26.11.2013, 11th Swiss Geoscience Meeting

A. Claude, N. Akcar, S. Ivy-Ochs, H.R. Graf, P.W. Kubik, C. Vockenhuber, A. Dehnert, M. Rahn and C. Schlüchter
Cosmogenic nuclide dating of Swiss Deckenschotter
France, Paris, 27.-31.08.2013, 8th IAG International Conference on Geomorphology

N. Dannhaus, F. von Blanckenburg, H. Wittmann, P. Kram and M. Christl
Measuring denudation rates with the $^{10}Be_{meteoric}/^9Be$ isotope ratio in catchments with different lithologies
Italy, Florence, 25.-30.08.2013, Goldschmidt 2013 Conference

A. Daraoui, M. Raiwa, M. Schwinger, M. Gorny, B. Riebe, C. Walther, M. Christl, C. Vockenhuber and H.-A. Synal
Iod-129 im Nordpolarmeer
Germany, Darmstadt, 01.-04.09.2013, Gesellschaft Deutscher Chemiker

P. Deline, N. Akçar, S. Ivy-Ochs and P.W. Kubik
Repeated rock avalanches onto the Brenva Glacier (Mont Blanc massif, Italy) during the Holocene
France, Paris, 27.-31.08.2013, 8th IAG International Conference on Geomorphology

P. Deline, N. Akçar, S. Ivy-Ochs and P.W. Kubik
Surface exposure dating with cosmogenic ^{10}Be of Late Holocene rock avalanches onto glaciers in the Mont Blanc massif, Italy
Austria, Vienna, 07.-12.04.2013, European Geosciences Union General Assembly 2013

M. Döbeli
Accelerator SIMS - dinosaurs in mass spectrometry?
Germany, Potsdam, 20.08.2013, Seventh Biennial Geochemical SIMS Workshop

M. Döbeli
Quantitative Oberflächenanalytik mit hochenergetischen Ionenstrahlen
Switzerland, Dübendorf, 29.10.2013, EMPATechnology Briefing 'Materialanalytik'

M. Döbeli, A. Dommann, X. Maeder, A. Neels, D. Passerone, H. Rudigier, D. Scopece, B. Widrig and J. Ramm
Surface layer evolution caused by the bombardment with ionized metal vapour
USA, Seattle, 25.06.2013, International Conference on Ion Beam Analysis

R. Grischott, F. Kober, S. Willett, K. Hippe and M. Christl
Paleo-denudation rates and possible links with climate variations in the Alps
Austria, Vienna, 07.-12.04.2013, European Geosciences Union Gerneral Assembly 2013

R. Grischott, F. Kober, K. Hippe, M. Lupker, M. Christl, I. Hajdas and S. Willett
Paleodenudation rates and possible links with climate and sediment flux variations in the Engadine, Eastern Swiss Alps
Switzerland, Lausanne, 16.11.2013, 11th Swiss Geoscience Meeting

D. Güttler, J. Beer, G. Boswijk, A. G. Hogg, J.G. Palmer, C. Vockenhuber, L. Wacker, J. Wunder
Worldwide detection of the rapid increase of cosmogenic ^{14}C 775AD
Portugal, Lisbon, 23.05.2013, Nuclear Physics in Astrophysics

I. Hajdas
Radiocarbon dating of 'old' and 'young'—an overview of applications in archaeology and art
Belgium, Brussels, 06.-08.06.2013, ArtConnoisseurs: A rendezvous with Art, Knowledge and Beauty

I. Hajdas, K. Hippe, and M. Maurer
Radiocarbon dating of bones at the ETH laboratory – an overview of methods and problems
France, Lyon, 22.-25.10.2013, 7th International Bone Diagenesis Meeting

I. Hajdas and C. Cristi
^{14}C on antique textiles -- New perspectives from nearly two decades of experience
Belgium, Ghent, 08.-12.04.13, Radiocarbon and Archeology

I. Hajdas
Radiocarbon dating and other cosmogenic isotopes with AMS technique.
Poland, Warsaw, 13.-15.03.2013, IsoG 2013 Environmental geochemistry: methods, trends, questions

I. Hajdas
^{14}C calibration curve update: IntCal13 (IntCal Working Group, in Prep)
Switzerland, Zürich, 29.05.2013, AMS Seminar

I. Hajdas
Absolute dating with ^{14}C : fascinating archives and samples
Switzerland, Berne, 03.05.2013, Opening workshop of Berne MICADAS

B. Hattendorf, C. Münsterer, R. Dietiker, J. Koch, L. Wacker, M. Christl, H.-A. Synal and D. Günther
Laser ablation for spatially resolved radiocarbon measurements with gas source-Accelerator Mass Spectrometry
Italy, Florence, 25.-30.08.2013, Goldschmidt 2013 Conference

K. Hippe, F. Kober, S. Ivy-Ochs, M. Lupker, L. Wacker, M. Christl and R. Wieler
Complex in situ cosmogenic ^{10}Be-^{14}C data suggest mid-Holocene climate change on the Bolivian Altiplano
Italy, Florence, 25.-30.08.2013, Goldschmidt 2013 Conference

K. Hippe, I. Hajdas and S. Ivy-Ochs
Timing of Marine Isotope Stage 3 (MIS 3) climate change recorded in Switzerland
Scottland, Blair Atholl, 28.-30.04.2013, COST-INTIMATE Annual Spring Workshop

K. Hippe, I. Hajdas and S. Ivy-Ochs
Radiocarbon chronologies of 45-20 ka BP - New results from the TiMIS Project
Austria, Obergurgl, 03.-07.10.2013, INQUA-INTIMATE Alpine Quaternary Workshop

K. Hippe, I. Hajdas, S. Ivy-Ochs, and M. Maisch
Middle Würm radiocarbon chronologies in the Swiss Alpine foreland - first results from the TiMIS project
Switzerland, Lausanne, 15.-26.11.2013, 11th Swiss Geoscience Meeting

S. Ivy-Ochs
Summary of cosmogenic nuclide exposure ages for the Alps
Austria, Obergurgl, 05.10.2013, Alpine Quaternary Workshop

S. Ivy-Ochs
Cosmogenic nuclides and the dating of Lateglacial moraines
Scotland, Aberdeen, 14.11.2013, Younger Dryas Leverhulme Network

R.S. Jones, A.N. Mackintosh, K.P. Norton, N.R. Golledge and P.W. Kubik
The glaciology of a Transantarctic Mountains outlet glacier – Implications for cosmogenic nuclide dating
New Zealand, Dunedin, 12.2.2013, Snow and Ice Research Group Meeting

R.S. Jones, A.N. Mackintosh, N.R. Golledge, K.P. Norton and P.W. Kubik
The glaciology of a Transantarctic Mountains outlet glacier: Implications for cosmogenic nuclide dating
New Zealand, Lower Hutt, 25.2.2013, Joint Antarctic Research Institute-Past Antarctic Climate Symposium

G. Klobe
Phasenraum und Energieverteilung aus der myCADAS-Quelle
Switzerland, Zurich, 18.09.2013, AMS Seminar

F. Kober, K. Hippe, B. Salcher, R. Grischott, M. Christl and N. Hählen
Sediment tracing, mixing and budgets in debris flow catchments: a cosmogenic nuclide perspective
Austria, Vienna, 07.-12.04.2013, European Geosciences Union General Assembly 2013

F. Kober, G. Zeilinger, K. Hippe, R. Grischott, T. Lendzioch, R. Zola and M. Christl
Sediment transfer and denudation rates across the Central Andes in Bolivia
Austria, Vienna, 07.-12.04.2013, European Geosciences Union General Assembly 2013

J. Lachner, M. Christl, S. Maxeiner, A. M. Müller, M. Suter and H.-A. Synal
Advances in low-energy AMS of ^{10}Be and ^{26}Al
Belgium, Namur, 12.09.2013, ECAART 11 Conference

J. Lachner, M. Christl, A. Müller, M. Suter and H.-A. Synal
Absorbermethode zur ^{10}Be- und ^{26}Al-Detektion
Germany, Hannover, 20.03.2013, DPG Frühjahrstagung

X. Mäder, A. Neels, A. Dommann, M. Döbeli, D. Passerone, D. Scopece, B. Widrig and J. Ramm
Design of high temperature corrosion resistant Al-Cr-O coatings deposited by arc operation for automotive applications
Switzerland, Basel, 21.05.2013, Swiss NanoConvention

S. K. Mandal, M. Lupker, J.-P. Burg, N. Haghipour and M. Christl
Low denudation recorded by ^{10}Be in river sands from Southern Peninsular, India
Switzerland, Lausanne, 16.11.2013, 11th Swiss Geoscience Meeting

S. Maxeiner, M. Christl, A. Herrmann, L. Wacker and H.-A. Synal
Current stripper designs and possible improvements - simulations and measurements
Switzerland, Zürich, 25.09.2013, AMS Seminar

M. Messerli, M. Maisch and S. Ivy-Ochs
GIS-based geomorphological mapping, dating of selected landforms and landscape evolution during the Lateglacial and Holocene, in the region of Val Tuoi, Grisons, Switzerland
France, Paris, 27.-31.08.2013, 8th IAG International Conference on Geomorphology

N. Mozafari, D. Tikhomirov, Ç. Özkaymak, Ö. Sümer, S. Ivy-Ochs, C. Vockenhuber, B. Uzel, H. Sözbilir and N. Akçar
Using cosmogenic ^{36}Cl to determine periods of enhanced seismicity in western Anatolia, Turkey
Switzerland, Lausanne, 15.-26.11.2013, 11th Swiss Geoscience Meeting

A.M. Müller
GIC - Basics and developments
Italy, Lecce, 01.03.2013, AMS seminar

A.M. Müller and M. Döbeli
Understanding and designing gas ionization chambers (GIC) with MAGICS
Slovenia, Ljubljana, 11.04.2013, SPIRIT JRA Meeting

A.M. Müller and M. Döbeli
MAGICS – A Mathematica® based gas ionization chamber simulation
Belgium, Namur, 11.09.2013, ECAART 11 Conference

A.M. Müller, M. Döbeli, M. Schulte-Borchers and M. Simon
Results and experiences with the CAEN DT5724 digitizer
Belgium, Namur, 12.09.2013, ECAART 11 Conference

C. Münsterer, B. Hattendorf, L. Wacker, J. Koch, M.Christl, R. Dietiker, H.-A. Synal and D. Günther
Ortsaufgelöste online ^{14}C-Analysen von Karbonaten mittels Laserablation-AMS
Germany, Hannover, 20.03.2013, DPG Frühjahrstagung

C. Münsterer, L. Wacker, B. Hattendorf, M.Christl, J. Koch, R. Dietiker, H.-A. Synal and D. Günther
^{14}C analysis of carbonates at high spatial resolution by the coupling of Laser Ablation with AMS
Switzerland, Bern, 26.-28.08.2013, Conference on Isotopes of Carbon, Water and Geotracers in Paleoclimate Research

C. Münsterer, L. Wacker, B. Hattendorf, M.Christl, J. Koch, R. Dietiker, H.-A. Synal and D. Günther
^{14}C analysis of carbonates at high spatial resolution by the coupling of Laser Ablation with AMS
Switzerland, Lausanne, 06.09.2013, SCS Fall Meeting 2013

C. Münsterer, L. Wacker, B. Hattendorf, M.Christl, J. Koch, R. Dietiker, H.-A. Synal and D. Günther
^{14}C analysis of carbonates at high spatial resolution by coupling laser ablation with Accelerator Mass Spectrometry
Switzerland, Interlaken, 29.-30.11.2013, Chanalysis 2013

J. E. Olson, D. D. Jenson, T. E. Lister, M. L. Adamic, M. G. Watrous and C. Vockenhuber
Alternative method for iodine AMS target preparation
Belgium, Namur, 12.09.2013, ECAART 11 Conference

M. Passarge
Phasenraum-Messungen am myCADAS
Switzerland, Zurich, 30.04.2013, AMS Seminar

B. Pichler, D. Brandova, C. Alewell, S. Ivy-Ochs, P.W. Kubik, C. Kneisel, K. Meusburger, M. Ketterer and M. Egli
The effect of permafrost on soil erosion using meteoric ^{10}Be, ^{137}Cs and $^{239,240}Pu$ in the Eastern Swiss Alps
Austria, Vienna, 07.-12.04.2013, European Geosciences Union Gerneral Assembly 2013

R. Reber, D. Tikhomirov, N. Akçar, S. Yesilyurt, V. Yavuz, P.W. Kubik and C. Schlüchter
Late Quaternary glacial history in northeastern Anatolia
Switzerland, Lausanne, 15.-26.11.2013, 11th Swiss Geoscience Meeting

R. Reber, D. Tikhomirov, N. Akçar, S. Yesilyurt, V. Yavuz, P.W. Kubik and C. Schlüchter
Late Pleistocene glacier advances in North Anatolia deduced from cosmogenic ^{10}Be and ^{26}Al
France, Paris, 27.-31.08.2013, 8th IAG International Conference on Geomorphology

R. Reber, D. Tikhomirov, N. Akcar, S. Yesilyurt, V. Yavuz, P.W. Kubik and C. Schlüchter
Paleoglaciations in northeastern Anatolia and their paleoenvironmental implications
Georgia, Tiflis, 12.-19.10.2013, IGCP 610 Meeting

B. Scherrer, H. Galinski, M. Döbeli and J. Cairney
Oxygen transport in noble metals
USA, Boston, 03.06.2013, MRS Fall Meeting

I. Schimmelpfennig, J. Schäfer, N. Akçar, S. Ivy-Ochs, R. Finkel, S. Zimmerman and C. Schlüchter
High-precision ^{10}Be-dating and Little Ice Age glacier advances at Steingletscher (Swiss Alps)
Italy, Florence, 25.-30.08.2013, Goldschmidt 2013 Conference

S. Schneider, M. Christl, R. Michel, G. Steinhauser and C. Walther
Bestimmung des Pu-Isotopenverhältnis in Proben aus Fukushima mittels AMS und RIMS
Germany, Hannover, 20.03.2013, DPG Frühjahrstagung

M. Schulte-Borchers, A. Eggenberger, M. Döbeli, M.J. Simon, A.M. Müller and H.-A. Synal
Heavy ion microbeams extracted into air using a capillary microprobe
USA, Seattle, 23.-28.06.2013, International Conference on Ion Beam Analysis

M.Schulte-Borchers, M. Döbeli, A. Eggenberger, M.J. Simon, A.M. Müller and H.-A. Synal
MeV SIMS with a capillary microprobe
Italy, Trieste, 30.09.-04.10.2013, Joint ICTP-IAEA Workshop on "Advanced Ion Beam Techniques: Imaging and Characterisation with MeV ions"

M. Schulte-Borchers
MeV SIMS using a glass capillary
Switzerland, Zurich, 16.10.2013, AMS Seminar

M. Schulte-Borchers
Generating heavy ion microbeams with a capillary microprobe for a future MeV SIMS setup
Switzerland, Zurich, 29.-30.08.2013, Zurich PhD seminar

M. Seiler, S. Maxeiner, M. Passarge and H.-A. Synal
myCADAS - Radiocarbon MS
Germany, Hannover, 20.03.2013, DPG Frühjahrstagung

M. Seiler
myCADAS - results and technical developements
Switzerland, Zürich, 23.10.2013, AMS Seminar

M. Suter
Wie optimiert man ein AMS system
Germany, Köln, 17.10.2013, Seminar Universität Köln

M. Suter, M. Christl, J. Lachner, S. Maxeiner, A.M. Müller, M. Seiler, H.A. Synal, C. Vockenhuber and L.Wacker
Challenges in the develpment of compact AMS facilities
Australia, Canberra, 12.04.2013, HIAS Conference 2013

H.-A. Synal
Latest developments in Accelerator Mass Spectrometry: Overview of modern facilities and research programs
India, New Delhi, 04.04.2013, Conference on Particle Accelerators: Technology and Applications in Science

H.-A. Synal
From the Oeschger counter to MICADAS: Latest developments in Accelerator Mass Spectrometry
Switzerland, Berne, 03.05.2013, Opening workshop of Berne MICADAS

H.-A. Synal
Progress in ^{14}C AMS
Germany, Bremerhafen, 28.11.2013, Workshop on Options of AMS in Earth Sciences

H.-A. Synal
Mass spectrometric C-14 detection techniques: Progress report
USA, San Francisco, 13.12.2013, AGU Fall Meeting

D. Tikhomirov, N. Akçar, S. Ivy-Ochs, and C. Schlüchter
Advanced model for limestone fault scarp dating and paleoearthquake history reconstruction
Switzerland, Lausanne, 15.-26.11.2013, 11th Swiss Geoscience Meeting

V. Vanacker, N. Bellin and P.W. Kubik
Human-induced changes in geomorphic process rates: Can we gain new insights when analysing magnitude-frequency distributions?
France, Paris, 27.-31.08.2013, 8th IAG International Conference on Geomorphology

V. Vanacker, A. Molina, P. Borja, G. Govers and P.W. Kubik
Managing erosion and sediment in rural watersheds
Belgium, Brussels, 11.10.2013, General Assembly of theBelgian committee of UNESCO-IHP

C. Vockenhuber
AMS and Accelerator-SIMS for astrophysics
Portugal, Lisbon, 23.05.2013, Nuclear Physics in Astrophysics

C. Vockenhuber
AMS of stable isotopes - shouldn't that be easy?
Austria, Vienna, 14.11.2013, VERA Seminar

C. Vockenhuber
Energy-loss straggling measurements
Switzerland, Zurich, 15.05.2013, AMS Seminar

L. Wacker, D. Güttler, H.-A. Synal and N. Walti
Precise radiocarbon dating by means of rapid atmospheric radiocarbon changes
Belgium, Ghent, 08.-12.04.13, Radiocarbon and Archeology

L. Wacker, J. Bourquin, S. Fahrni, C. McIntyre and H.-A. Synal
Gas ion source and automated graphitization equipment
Germany, Bremerhaven, 28.11.2013, Seminar Alfred Wegener Institut

A. Wallner, R. Capote Noy, M. Christl, I. Dillmann, L.K. Fifield, F. Käppeler, A. Klix, A. Krása, C. Lederer,
J. Lippold, A. Plompen, V. Semkova, M. Srncik, P. Steier, S. Tims and S. Winkler
Neutron-induced reactions on U and Th – a new approach via AMS
Belgium, Geel, 05.-08.11.2013, NEMEA-7/CIELO International Collaboration on Nuclear Data

A. Wallner, R. Capote, M. Christl, K. Fifield, M. Srncik, S. Tims, F. Käppeler, A. Krasa, A. Plompen,
J. Lippold, P. Steier and S. Winkler
Neutron-induced reactions on U and Th - New cross-section measurements via AMS
USA, New York, 06.03.2013, 2013 International Nuclear Data Conference for Science and Technology

C. Wirsig, S. Ivy-Ochs, J. Zasadni, N. Akcar, M. Christl and C. Schlüchter
Towards a 'true' age of LGM ice surface decay in Oberhaslital
Switzerland, Lausanne, 15.-26.11.2013, 11th Swiss Geoscience Meeting

C. Wirsig, S. Ivy-Ochs, J. Zasadni, N. Akcar, M. Christl, P.W. Kubik and C. Schlüchter
Datierung des Zerfalls der Eismasse des letzteiszeitlichen Maximums in den Alpen mit kosmogenem ^{10}Be
Germany, Hannover, 20.03.2013, DPG Frühjahrstagung

C. Wirsig, S. Ivy-Ochs, J. Zasadni, N. Akcar, M. Christl and C. Schlüchter
When did the LGM ice surface decay in the high Alps?
Switzerland, Bern, 28.03.2013, Exogene Geology and Quaternary Global Change Seminar

C. Wirsig, S. Ivy-Ochs, J. Zasadni, N. Akcar, M. Christl, P. Deline and C. Schlüchter
Timing of ice decay after the LGM in the high Alps
France, Paris, 28.08.2013, IAG International Conference on Geomorphology

C. Wirsig, S. Ivy-Ochs, J. Zasadni, N. Akcar, M. Christl, P. Deline and C. Schlüchter
When did the LGM ice surface decay in the high Alps?
Switzerland, Zürich, 10.04.2013, Current topics from Accelerator Mass Spectrometry and its applications

C. Wirsig, S. Ivy-Ochs, J. Zasadni, N. Akcar, M. Christl and C. Schlüchter
Timing of LGM ice downwasting across the Alps
Switzerland, Zürich, 11.12.2013, INTIMATE workshop

L. Wüthrich, R. Zech, N. Haghipour, C. Gnägi, M. Christl and S. Ivy-Ochs
Dating glacial deposits in the western Swiss lowlands using cosmogenic ^{10}Be
Switzerland, Lausanne, 15.-26.11.2013, 11th Swiss Geoscience Meeting

G. Zeilinger, F. Kober, K. Hippe, T. Lenzioch, R. Grischott, R. Pillco Zolá and M. Christl
Tectonic control on denudation rates in the central Bolivian Andes
Austria, Vienna, 07.-12.04.2013, European Geosciences Union Gerneral Assembly 2013

SEMINAR
'CURRENT TOPICS IN ACCELERATOR MASS SPEKTRO- METRY AND RELATED APPLICATIONS'

Spring semester

20.02.2013

Stephanie Schneider (IRS Hannover, Germany), Untersuchung von Bodenproben aus Fukushima in Bezug auf Pu-Isotopenverteilung mittels AMS

27.02.2013

Georg Rugel (DREAMS, Dresden, Germany), Status of the new DREsden AMS facility - DREAMS

06.03.2013

Romain Delunel (University of Bern, Switzerland), Cosmic-ray flux attenuation by seasonal snow cover revealed by neutron-detector monitoring: Potential implications for cosmic-ray exposure dating

13.03.2013

Sara Azimi (National University of Singapore, Singapore), Three-dimensional silicon micro- and nano-machining

13.03.2013

Nicolas Gruber (ETHZ, Switzerland), Oceanic sources and sinks for atmospheric CO_2: recent trends and insights

19.03.2013

Samuel Haas (ETHZ, Switzerland), U-236 Messungen bei 200 kV

27.03.2013

Kristina Hippe (ETHZ, Switzerland), Late Holocene climate change on the Bolivian Altiplano: Evidence from complex ^{10}Be and in situ ^{14}C data

03.04.2013

Gideon Henderson (University of Oxford, United Kingdom), Using radionuclides to quantify trace-metal cycles in the ocean

10.04.2013

Christian Wirsig (ETHZ, Switzerland), ^{10}Be surface exposure dating of the LGM ice surface decay in the high Alps

17.04.2013

Andreas Eggenberger (ETHZ, Switzerland), Angular distribution induced by capillary focusing of ion beams in air

25.04.2013

Florian Adolphi (Lund University, Sweden), Solar activity variations at the end of the last ice age, and implications for solar forcing of climate

30.04.2013
Michelle Passarge (ETHZ, Switzerland), Phasenraum-Messungen am MyCADAS

08.05.2013
Susan Trumbore (MPI Biogeochemistry, Jena, Germany), Age of carbon respired from ecosystems using radiocarbon

15.05.2013
Christof Vockenhuber (ETHZ, Switzerland), Energy loss straggling measurements

22.05.2013
Maarten Lupker (ETHZ, Switzerland), Depth-dependence of the production rate of in-situ ^{14}C in quartz

29.05.2013
Irka Hajdas (ETHZ, Switzerland), Update of the radiocarbon calibration curve

Fall semester

28.08.2013
Elena Chamizo (CNA, Seville, Spain), AMS at the Centro Nacional des Acceleradores (CNA) – History and perspectives

17.09.2013
Markus Thöni (ETHZ, Switzerland), Energy-loss straggling measurements in gases

18.09.2013
Gunther Klobe (ETHZ, Switzerland), Phasenraum und Energieverteilung aus der MyCADAS-Quelle

25.09.2013
Sascha Maxeiner (ETHZ, Switzerland), Current stripper designs and possible improvements - simulations and measurements

01.10.2013
Anton Wallner (ANU, Canberra, Australia), AMS studies of neutron-induced reactions on actinides - crucial data for nuclear technology

02.10.2013
Hervé Bocherens (University of Tübingen, Germany), Stable isotopes in Quaternary mammals: Tools for reconstructing palaeoenvironments and palaeodiets

09.10.2013
Thomas Blattmann (ETHZ, Switzerland), A source to sink carbon isotope tracer study of the Lake Constance catchment and basin

16.10.2013
Martina Schulte-Borchers (ETHZ, Switzerland), Using a glass capillary for heavy ion microbeams and MeV SIMS

23.10.2013

Martin Seiler (ETHZ, Switzerland), MyCADAS - results and technical developments

24.10.2013

Xiongxin Dai (AECL CRL, Chalk River, Canada), Actinides by AMS

30.10.2013

Markus Egli (UZH, Switzerland), Soil formation and erosion in Alpine areas: how can we assess time-split rates?

06.11.2013

Nadia Walti (ETHZ, Switzerland), Messung und Modellierung des 775 AD Radiokarbon Ereignisses

13.11.2013

Peter Zotter (PSI, Switzerland), Radiocarbon-based source apportionment of carbonaceous aerosols

20.11.2013

Tobias Lorenz (PSI, Switzerland), Determination of radionuclides from spallation targets

27.11.2013

Detlef Günther (ETHZ, Switzerland), Advanced and New Capabilities Using Laser Ablation Inductively Coupled Plasma Mass Spectrometry

04.12.2013

Reto Grischott (ETHZ, Switzerland), Holocene paleo-denudation rates in the Alps derived by cosmogenic nuclides

11.12.2013

Nuria Casacuberta (ETHZ, Switzerland), The distribution of U-236, I-129, and Pu-isotopes in the Arctic Ocean

18.12.2013

Michael Franzmann (University of Mainz, Germany), Resonance ionization mass spectrometry (RIMS) on actinides: Analytical and spectroscopic applications

THESES (INTERNAL)

Term papers

Samuel Haas
Uran-236 Messungen bei niedrigen Energien und Detektor-Auflösungs-Messungen mit schweren Ionen
ETH Zurich (Switzerland)

Gunther Klobe
Phasenraumbestimmung bei einer Cs-Sputter-Ionenquelle am myCADAS
ETH Zurich (Switzerland)

Michelle Passarge
Phasenraum-Messungen am myCADAS
ETH Zurich (Switzerland)

Diploma/Master theses

Andreas Eggenberger
Angular Distribution induced by Capillary Focusing of Ion Beams in Air
ETH Zurich (Switzerland)

Nadia Walti
^{14}C Altersbestimmung mit Hilfe der Feinstruktur der Kalibrierkurve sowie Simulation der Radiokarbon-Produktion mit MATLAB Box-Modell des globalen, präindustriellen Kohlenstoffzyklus
ETH Zurich (Switzerland)

Doctoral theses

Johannes Lachner
Environmental applications of low-energy Accelerator Mass Spectrometry
ETH Zurich (Switzerland)

Marius Simon
Physikalische Grundlagen und Anwendungen einer Kapillaren-Mikrosonde
ETH Zurich (Switzerland)

THESES (EXTERNAL)

Term papers

Manuel Raiwa
Analyse von Iod-129 in Atlantikwasser
University of Hannover (Germany)

Diploma/Master theses

Mathias Bichler
Landscape evolution north of the Sonnblick (Salzburg) during the last 21 ka
University of Vienna (Austria)

Björn Dittmann
Bestimmung des Verteilungskoeffizienten von Plutonium(VI) an Kaolinit und Opalinuston im Ultraspurenbereich
University of Cologne (Germany)

Michael Imhof
Sputtered binary chromium alloys on polycarbonate substrates
University of South Australia (Australia)

David Meier
Elektrochemisch deponierte TiO_2 Schichten auf Titansubstrat
ETH Zurich (Switzerland)

Martin Meindl
Landscape evolution north of the Sonnblick (Salzburg) during the last 21 ka
University of Vienna (Austria)

Maja Messerli
GIS-based geomorphological mapping, dating of selected landforms and landscape evolution during the Lateglacial and Holocene, in the region of Val Tuoi, Grisons, Switzerland
University of Zurich (Switzerland)

Jan Nagelisen
Prehistoric Rock Avalanches in the Obersee Area, Glarner Alps, Switzerland
ETH Zurich (Switzerland)

Stephanie Schneider
Untersuchung von Bodenproben aus Fukushima in Bezug auf Pu Isotopenverteilung mittels AMS
University of Hannover (Germany)

Lorenz Wuthrich
Dating glacial deposits in the western Swiss lowlands using cosmogenic ^{10}Be
ETH Zurich (Switzerland)

Doctoral theses

Matthias Bator
Investigation of the magnetic and magnetoelectric properties of orthorhombic REMnO₃ thin films
ETH Zurich (Switzerland)

Nicolas Bellin
Human impact on soil degradation based on ^{10}Be cosmogenic radionuclide measurements in a semi-arid region: the case of Southeast Spain
Université Catholique de Louvain (Belgium)

Dorota Flak
New metal oxide nanoparticles for gas sensors
University of Krakov (Poland)

Angela Furrer
Colours in thin metallic films based on precious metals and their intermetallic phases
ETH Zurich (Switzerland)

Endre F. Gjermundsen
Quaternary glacial history of norhtern Spitsbergen, Svalbard; cosmogenic nuclide constraints on configuration, chronology and ice dynamics
The University Centre in Svalbard / University of Oslo (Norway)

Marie Guns
Sediment dynamics in tropical mountain regions: influence of anthropogenic disturbances on sediment transfer mechanisms
Université Catholique de Louvain (Belgium)

Negar Haghipour
Active deformation and landscape evolution of the Makran accretionary wedge (SE-Iran)
New constraints from surface exposure dating of fluvial terraces
ETH Zurich (Switzerland)

Yi Hu
Preparation and structural analysis of multiferroic rare earth manganate thin films
ETH Zurich (Switzerland)

Edvinas Navickas
Ion transport at hetero and homophase boundaries of YSZ thin films
University of Vienna (Austria)

Dieter Stender
Controlling strain and micro-structure of yttria stabilized zirconia thin films grown by pulsed laser deposition
ETH Zurich (Switzerland)

COLLABORATIONS

Australia

The Australian National University, Department of Nuclear Physics, Canberra

University of Sidney, Australian Centre for Microscopy & Microanalysis, Sidney

Austria

AlpS - Zentrum für Naturgefahren- und Riskomanagement GmbH, Geology and Mass Movements, Innsbruck

Geological Survey of Austria, Sediment Geology, Vienna

University of Innsbruck, Institute of Botany, Innsbruck

University of Innsbruck, Institute of Geography, Innsbruck

University of Innsbruck, Institute of Geology, Innsbruck

University of Vienna, VERA, Faculty of Physics, Vienna

Vienna University of Technology, Institute for Geology, Vienna

Belgium

Royal Institute for Cultural Heritage, Brussels

Université catholique de Louvain, Earth and Life Institute, Louvain-la-Neuve

Canada

Chalk River Laboratories, Dosimetry Services, Chalk River

ISOTRACE, Department of Physics, Ottawa

TRIUMF, Vancouver

China

China Earthquake Administration, Beijing

Peking University, Accelerator Mass Spectrometry Laboratory, Beijing

Denmark

Danfysik, A/S, Taastrup

Risø DTU, Risø National Laboratory for Sustainable Energy, Roskilde

University of Southern Denmark, Department of Physics, Chemistry and Pharmacy, Odense

Finnland

University of Jyväskylä, Physics Department, Jyväskylä

France

Aix-Marseille University, Collège de France, Aix-en-Provence

Laboratoire CRIIRAD, Valence

Laboratoire de Biogeochimie Moléculaire, Strasbourg

Université de Savoie, Laboratoire EDYTEM, Le Bourget du Lac

Université Paris Panthéon-Sorbonne,Laboratoire de Géographie Physique, Meudon cedex

Germany

Alfred Wegener Institute of Polar and Marine Research, Bremerhaven

BSH Hamburg, Radionuclide Section, Hamburg

Bundesamt für Strahlenschutz, Strahlenschutz und Umwelt, Neuherberg

Deutsches Bergbau Museum, Bochum

Diözesanmuseum Freising, Erzbischöfliches Ordinariat München, Freising

German Research Centre for Geosciences (GFZ), Potsdam

Hydroisotop GmbH, Schweitenkirchen

Leibniz-Institut für Ostseeforschung Warnemünde, Marine Geologie, Rostock

Regierungspräsidium Stuttgart, Landesamt für Denkmalpflege, Esslingen

University of Cologne, Institute of Geology and Mineralogy, Cologne

University of Cologne, Physics Department, Cologne

University of Freiburg, Institut für Vorderasiatische Archäologie, Freiburg

University of Hannover, Institute for Radiation Protection and Radioecology, Hannover

University of Heidelberg, Institute of Environmental Physics, Heidelberg

University of Hohenheim, Institute of Botany, Stuttgart

University of Münster, Institute of Geology and Paleontology, Münster

University of Tübingen, Department of Geosciences, Tübingen

Direktion Landesarchäologie, Speyer

Hungary

Hungarian Academy of Science, Institute of Nuclear Research (ATOMKI), Debrecen

India

Inter-University Accelerator Center, Accelerator Division, New Dehli

Israel

Hebrew University, Geophysical Institute of Israel, Jerusalem

Italy

CNR Rome, Institute of Geology, Rome

Geological Survey of the Provincia Autonoma di Trento, Landslide Monitoring, Trentino

INGV Istituto Nazionale di Geofisica e Vulcanologia, Sez. Sismologia e Tettonofisica, Rome

University of Bologna, Earth Sciences, Bologna

University of Padua, Department of Geology and Geophysics, Padua

University of Salento, Department of Physics, Lecce

University of Turin, Department of Earth Sciences, Turin

Liechtenstein

OC Oerlikon AG, Balzers

Mexico

UNAM (Universidad Nacional Autonoma de Mexico), Instituto de Fisica, Mexico

Monaco

IAEA Environment Laboratories, Department of Nuclear Sciences and Applications, Monaco

New Zealand

Victoria University of Wellington, School of Geography, Environment and Earth Sciences, Wellington

Norway

The University Centre of Svalbard, Quaternary Geology, Longyearbyen

University of Bergen, Department of Biology, Bergen

University of Bergen, Department of Earth Science, Bergen

University of Norway, The Bjerkness Centre for Climate Res., Bergen

Poland

University of Marie Curie Sklodowska, Department of Geography, Lublin

Romania

Horia Hulubei - National Institute for Physics and Nuclear Engineering, Magurele

Russia

Russian Academy of Sciences, Laboratory of Ion and Molecular Physics of the Institute for Energy Problems of Chemical Physics, Moscow

Singapore

National University of Singapore, Department of Chemistry, Singapore

Slovakia

Comenius University, Faculty of Mathematics, Physics and Infomatics, Bratislava

Slovenia

Geological Survey of Slovenia, Ljubljana

Spain

Autonomous University of Barcelona, Environmental Science and Technology Institute, Barcelona

University of Murcia, Department of Plant Biology, Murcia

University of Seville, Department of Applied Physics, Seville

University of Seville, National Center for Accelerators, Seville

Sweden

Lund University, Department of Earth and Ecosystem Sciences, Lund

Onsala Space Obervatory, Onsala

University of Uppsala, Angström Institute, Uppsala

Uppsala University, Tandem Laboratory, Uppsala

Switzerland

Amt für Kultur Kanton Graubünden, Archäologischer Dienst, Chur

Dendrolabor Wallis, Brig

Empa, Advanced Materials Processing, Dübendorf

Empa, Functional Polymers, Dübendorf

Empa, Joining Technology and Corrosion, Dübendorf

Empa, Solid State Chemistry, Dübendorf

Empa, Thin Films, Dübendorf

Empa, Mechanics of Materials and Nanostructures, Thun

ENSI, Brugg

ETH Zurich, Department of Health Sciences and Technology, Zurich

ETH Zurich, Engineering Geology, Zurich

ETH Zurich, Institute of Electrical Engineering, Zurich

ETH Zurich, Institute of Geology, Zurich

ETH Zurich, Institute of Isotope Geochemistry and Mineral Resources, Zurich

ETH Zurich, Laboratory of Inorganic Chemistry, Zürich

ETH Zurich, Metals Research, MATL, Zurich

Evatec AG, Flums

Federal Office for Civil Protection, Spiez Laboratory, Spiez

Glas Trösch AG, Bützberg

Gübelin Gem Lab Ltd. (GGL), Luzern

Haute Ecole ARC, IONLAB, La-Chaux-de-Fonds

Helmut Fischer AG, Hünenberg

II-VI Laserenterprise, Zurich

Kanton Bern, Achäologischer Dienst, Berne

Kanton Graubünden, Archäologischer Dienst, Chur

Kanton Graubünden, Kantonsarchäologie, Chur

Kanton Solothurn, Kantonsarchäologie, Solothurn

Kanton St. Gallen, Kantonsarchäologie, St. Gallen

Kanton Zug, Kantonsarchäologie, Zug

Kanton Zürich, Kantonsarchäologie, Dübendorf

Labor für quartäre Hölzer, Affoltern a. Albis

Laboratiore Romand de Dendrochronologie, Moudon

Musée Romain d'Avenches, Restauration, Avenches

Office et Musée d'Archéologie Neuchatel, , Neuchatel

Paul Scherrer Institut (PSI), High Temperature Materials, Villigen

Paul Scherrer Institut (PSI), Laboratory for Atmospheric Chemistry, Villigen

Paul Scherrer Institut (PSI), Laboratory for Radiochemistry and Environmental Chemistry, Villigen

Paul Scherrer Institut (PSI), Materials Group, Villigen

Paul Scherrer Institut (PSI), Materials Science Beamline, Villigen

Paul Scherrer Institut (PSI), Muon Spin Rotation, Villigen

Research Station Agroscope Reckenholz-Tänikon ART, Air Pollution / Climate Group, Zurich

Stadt Zürich, Amt für Städtebau, Zurich

Swiss Federal Institute for Forest, Snow and Landscape Reseach (WSL), Landscape Dynamics, Dendroecology, Birmensdorf

Swiss Federal Institute for Forest, Snow and Landscape Reseach (WSL), Soil Sciences, Birmensdorf

Swiss Federal Institute of Aquatic Science and Technology (Eawag), Radioactive Tracers, Dübendorf

Swiss Gemmological Institute - SSEF, Basel

Swiss Institute for Art Research, SIK ISEA, Zurich

Universität Basel, Department Altertumswissenschaften, Basel

University of Basel, Department of Physics, Basel

University of Basel, Institut für Prähistorische und Naturwissenschaftliche Archäologie (IPNA), Basel

University of Bern, Climate and Environmental Physics, Berne

University of Bern, Department of Chemie and Biochemistry, Berne

University of Bern, Institute of Geology, Berne

University of Bern, Oeschger Center for Climate Research, Berne

University of Fribourg, Department of Physics, Fribourg

University of Geneva, Department of Anthropology and Ecology, Geneva

University of Geneva, Department of Geology and Paleontology, Geneva

University of Lausanne, Institute of Geomatics and Risk Analysis, Lausanne

University of Neuchatel, Department of Geology, Neuchatel

University of Zurich, Institute of Geography, Zurich

University of Zurich, Abteilung Ur- und Frühgeschichte, Zurich

University of Zurich, Paläontologisches Institut und Museum, Zurich

Turkey

Istanbul Technical University, Faculty of Mines, Istanbul

UK

University of Cambridge, Department of Earth Sciences, Cambridge

University of Oxford, Department of Earth Sciences, Oxford

United Kingdom

Durham University, Department of Geography, Durham

USA

Colorado State University, Department of Environmental and Radiological Health Sciences, Fort Collins

Columbia University, LDEO, New York

Eckert & Ziegler Vitalea Science, AMS Laboratory, Davis

Idaho National Laboratory, National and Homeland Security, Idaho Falls

University of Utah, Geology and Geophysics, Salt Lake City

Woods Hole Oceanographic Institution, Center for Marine and Environmental Radioactivity, Woods Hole

VISITORS AT THE LABORATORY

Andrew P. Moran
Institute of Geography, University of Innsbruck, Innsbruck, Austria
05.01.2013 to 04.02.2013

Tess Van den Brande
Royal Institute for Cultural Heritage, Brussels, Belgium
07.01.2013 to 11.01.2013

Jens Jensen
Department of Physics, Chemistry and Biology - IFM, Linköping University, Linköping, Sweden
16.01.2013

Orkhan Osmani
Faculty of Physics, University of Duisburg-Essen, Duisburg, Germany
16.01.2013

Peter Sigmund
Department of Physics, Chemistry and Pharmacy, University of Southern Denmark, Odense, Denmark
16.01.2013

Harry Withlow
IONLAB ARC, Haute Ecole Arc. Ingénierie, La Chaux-de-Fonds, Switzerland
16.01.2013

Natko Skukan
Laboratory for Ion Beam Interactions, Rudjer Boskovic Institute, Zagreb, Croatia
24.01.2013 to 25.01.2013

Mathieu Boudin
Royal Institute for Cultural Heritage, Brussels, Belgium
11.02.2013 to 15.02.2013

Bruno Luigi
Dottorato di Ricerca in Stratigrafia e Sedimentologia, Università degli Studi di Bologna, Bologna, Italy
15.02.2013 to 15.05.2013

Joel Rehmann
Kantonsschule Wettingen, Wettingen, Switzerland
18.02.2013 to 08.03.2013

Amine Cassimi
CEA/CNRS, Grand Accélérateur National d'Ions Lourds, Caen, France
19.02.2013 to 21.02.2013

Eduardo Jorge da Costa Alves
Ion Beam Laboratory, Unit of Physics and Accelerators, Instituto Superior Técnico (IST)/Instituto Tecnológico e Nuclear (ITN), Sacavém, Portugal
19.02.2013 to 20.02.2013

Florent Durantel
CEA/CNRS, Grand Accélérateur National d'Ions Lourds, Caen, France
19.02.2013 to 21.02.2013

Maria Micaela Fonseca
Ion Beam Laboratory, Unit of Physics and Accelerators, Instituto Superior Técnico (IST)/Instituto Tecnológico e Nuclear (ITN), Sacavém, Portugal
19.02.2013 to 20.02.2013

Stéphane Guillous
CEA/CNRS, Grand Accélérateur National d'Ions Lourds, Caen, France
19.02.2013 to 21.02.2013

Stephanie Schneider
Institute for Radioecology and Radiation Protection, Leibniz University of Hannover, Hannover, Germany
19.02.2013 to 20.02.2013

Tess Van den Brande
Royal Institute for Cultural Heritage, Brussels, Belgium
25.02.2013 to 01.03.2013

Georg Rugel
Helmholtz-Zentrum Dresden-Rossendorf, Dresden, Germany
27.02.2013

Milko Jakšić
Laboratory for Ion Beam Interactions, Rudjer Boskovic Institute, Zagreb, Croatia
04.03.2013 to 06.03.2013

Romain Delunel
Department of Geology, University of Bern, Berne, Switzerland
06.03.2013

Nadia Walti
Department of Chemistry, ETH Zurich, Zurich, Switzerland
11.03.2013 to 30.06.2013

Sara Azimi
Faculty of Science, Department of Physics, National University of Singapore, Singapore, Singapore
13.03.2013

Gideon Henderson
Department of Earth Science, University of Oxford, Oxford, United Kingdom
03.04.2013 to 04.04.2013

Pere Masque
Department of Physics, University of Barcelona, Barcelona, Spain
03.04.2013 to 04.04.2013

Michiel Rutgers van der Loeff
Alfred-Wegener-Institut, Bremerhaven, Germany
03.04.2013 to 04.04.2013

Peter Steier
Vienna Environmental Research Accelarator, University of Vienna, Vienna, Austria
03.04.2013 to 04.04.2013

Stephan Winkler
Vienna Environmental Research Accelarator, University of Vienna, Vienna, Austria
03.04.2013 to 04.04.2013

Florian Adophi
Department of Geology, Lund University, Lund, Sweden
07.04.2013 to 27.04.2013

Nicholas Priest
AECL Chalk River Laboratories, Chalk River, Canada
16.04.2013

Susan Trumbore
Max Planck Institute for Biogeochemistry, Jena, Germany
08.05.2013

Amit Roy
Inter-University Accelerator Centre, New Delhi, India
04.06.2013 to 06.06.2013

Conradie Lowry
iThemba LABS, Somerset West, South Africa
26.06.2013 to 27.06.2013

Young Graeme
Biotransformation & Drug Disposition (BDD), GlaxoSmithKline, Hertfordshire, United Kingdom
27.06.2013

Elena Chamizo
Centro Nacional de Aceleradores, University of Seville, Seville, Spain
01.08.2013 to 31.08.2013

Douglas D. Jenson
Idaho National Laboratory, Idaho Falls, USA
17.09.2013 to 18.09.2013

Andreas Moor
Institute for Radioecology and Radiation Protection, Leibniz University of Hannover, Hannover, Germany
17.09.2013 to 18.09.2013

Mareike Schwinger
Institute for Radioecology and Radiation Protection, Leibniz University of Hannover, Hannover, Germany
17.09.2013 to 18.09.2013

Victor Alarcon Diez
SAFIR, Université Pierre et Marie Curie, Paris, France
23.09.2013 to 25.09.2013

Carlo Tintori
CAEN S.p.A., Viareggio, Italy
23.09.2013 to 25.09.2013

Ian Vickridge
SAFIR, Université Pierre et Marie Curie, Paris, France
23.09.2013 to 25.09.2013

Robine Schiesser
Bezirksschule Wettingen, Wettingen, Switzerland
27.09.2013

Graham Entwistle
Isoprime Ltd., Cheadle, United Kingdom
30.09.2013 to 04.10.2013

Hervé Bocherens
Department of Geoscience, Eberhard Karls University of Tübingen, Tübingen, Germany
01.10.2013 to 02.10.2013

Anton Wallner
Department of Nuclear Physics, Australian National University, Canberra, Australia
01.10.2013

Satinath Gargari
Inter-University Accelerator Centre, New Delhi, India
15.10.2013 to 12.11.2013

Pankaj Kumar
Inter-University Accelerator Centre, New Delhi, India
15.10.2013 to 26.11.2013

Xiongxin Dai
AECL Chalk River Laboratories, Chalk River, Canada
24.10.2013 to 25.10.2013

Lukas Jakob
Kantonsschule Olten, Olten, Switzerland
18.11.2013 to 22.11.2013

Tynan Richards
Kantonsschule Olten, Olten, Switzerland
18.11.2013 to 22.11.2013

Yoann Fagault
Centre Européen de Recherche et d'Enseignement des Géosciences de l'Environnement, Aix-en-Provence, France
02.12.2013 to 13.12.2013

Thibaut Tuna
Centre Européen de Recherche et d'Enseignement des Géosciences de l'Environnement, , Aix-en-Provence, France
02.12.2013 to 13.12.2013